D0090421

WHY DON'T JUMBO JETS
FLAP THEIR WINGS?

WHY DON'T JUMBO JETS

Flap Their Wings?

Flying Animals,
Flying Machines,
and How They Are Different

David E. Alexander

RUTGERS UNIVERSITY PRESS

NEW BRUNSWICK, NEW JERSEY, AND LONDON

Library of Congress Cataloging-in-Publication Data

Alexander, David E., 1955–

Why don't jumbo jets flap their wings? : flying animals, flying machines, and how they are different / David

E. Alexander.

p. cm.

Includes bibliographical references and index.

ISBN 978–0–8135–4479–3 (hardcover : alk. paper)

1. Aeronautics—Popular works. 2. Animal flight—Popular works. 3. Flying-machines—Popular works. 4.

Airplanes—Popular works. 5. Birds—Flight—Popular works. I. Title.

TL546.7.A44 2009

629.13—dc22

2008035425

A British Cataloging-in-Publication record for this book is available from the British Library.

Text design by Adam B. Bohannon

Visit our Web site: http://rutgerspress.rutgers.edu

Manufactured in the United States of America

To Steve Vogel

 and to my parents,

  Eleanor Woolley and Edward Alexander

CONTENTS

ILLUSTRATIONS

PREFACE

Although flight symbolizes freedom to most people, I doubt if many dwell much on flight in the literal sense. Flight is going on all around us, however. It affects us in many ways, from the obvious, such as transportation, to the subtle, such as disease transmission and crop pollination. I suspect that most people, on seeing a jet airliner take off or a swan land, are casually, momentarily interested in flight. But once they learn a bit about flight, they may find that it is a naturally interesting subject.

I am one of those people who drops what he is doing and looks up when an airplane flies overhead, who nearly runs his car off the road while trying to drive and watch a large flock of migrating geese at the same time. I find the flight of airplanes and animals endlessly fascinating, and I am envious of both aerobatic aviators and the even more agile flyers of the animal kingdom. Can anyone watch the ponderous grace of a crane or a pelican and not be amazed at how easy they make their landings look? Can anyone watch a jumbo jet take off and not be amazed that thin air can hold up such an enormous contraption? I certainly can't, and I hope this book transmits some of my enthusiasm to you.

I grew up surrounded by aviation. My family lived near Dayton, Ohio (home of the Wright brothers), and my father worked at Wright-Patterson Air Force Base. My father and I both built display models of airplanes, and he, my brother, and I taught ourselves to fly gas-powered control-line airplane models. I learned photographic processing so I could print my own enlargements of photos of historic airplanes that I took at the U.S. Air Force Museum.

Later, as a professional biologist, I discovered that animal flight is just as fascinating as aviation and that the two share similarities and differences. I discovered that ordinary turkey vultures (sometimes known as buzzards), although clumsy and stunningly ugly on the ground, epitomize grace in the air. A buzzard's skill at working the tiny updrafts over a tree line to avoid flapping must be seen to be believed. Likewise, the wild aerial gyrations of a pair of dogfighting male dragonflies in a territorial dispute outdo Hollywood's best efforts, not to mention the Navy's real *Top Gun* school for fighter pilots. Flying across the ocean? For millennia, songbirds and shorebirds have migrated an oversea route twice each year, flying from New England to Venezuela in the fall and then back again in the spring. Flying on instruments in total darkness? For tens of millions of years, houseflies have possessed a built-in artificial horizon. Vertical takeoff and landing? These are routine actions for insects, bats, and all but the largest birds. Moving on the ground as easily as in the air (something that still eludes engineers despite almost a century of trying)? Nearly all insects and most birds can do so, and many can swim underwater as well.

Birds and airplanes face some of the same challenges: both need to be as light as possible, but neither can afford to be too fragile. The aerodynamics of their wings are surprisingly similar; although when small animals flap their wings very rapidly or airplanes fly near or above the speed of sound, the rules of the game diverge. In short, animals and airplanes are the same yet different, and I never tire of either. I can be perfectly happy to spend hours watching airplanes at an air show, crows playing in ridge lift at the Grand Canyon, pelicans fishing, or vintage airplanes sitting on the ground. The things that make them similar and different are both fascinating and form the subject of this book.

ACKNOWLEDGMENTS

Many people helped me track down information or answered my technical questions. They include Richard Colgren (airliner capabilities), Zack Falin (insect natural history and identification), Erik Floor (vertebrate nervous systems), John McMasters (aeronautical engineering), Charles Michener and Michael Engel (bees and insect evolution), Mark Robbins and Thor Holmes (anything ornithological), and Bob Timm (bats). Ron Barrett-Gonzalez patiently explained certain facets of sophisticated aerodynamics, helped me understand and deal with the reviewer's comments, and willingly answered my aerospace engineering questions. I thank all of them, but of course, any remaining errors are my responsibility.

I would like to thank David A. Reay for permission to use some of his drawings from *The History of Man-Powered Flight* (New York: Pergamon, 1977) and Paul Gladman, head picture librarian of the Flight Collection, for drawings of human-powered aircraft that first appeared in the journal *Flight International*. I originally intended to use these diagrams directly, but in the end I modified them substantially. I combined parts of two of Reay's drawings to form figure 10.1; and I redrew the complex cutaway drawings from *Flight International* into much simpler outline sketches, using one for figure 10.3 and combining two for figure 10.2. The other illustrations are my original drawings.

I also thank Moira Ozias, my friend and writing teacher. She read an early draft of the entire manuscript, made innumerable insightful and helpful suggestions, and asked some very thought-provoking questions, all of which vastly improved the text. My colleague and mentor, Steve Vogel, encouraged me to keep trying when the first few publishers turned down this manuscript. My son, Kevin Alexander, found several useful Internet sites and worked on early versions of some of the diagrams. My wife, Helen Alexander, is my number-one booster and went out of her way to help me find time to write. This included juggling job-related, professional, and child-care duties that, of course, resulted in more time but also less stress, more conducive conditions, and more useful consolidated blocks of time. I thank all of these wonderful people, as well as any others whom I may have forgotten.

WHY DON'T JUMBO JETS
FLAP THEIR WINGS?

~ 1 ~

Flying Animals
and Flying Machines
Birds of a Feather?

A irplanes fly, birds fly, insects fly, helicopters fly. I wish I could fly; but without elaborate mechanical contrivances, I can't. What lets a pigeon or even a lowly mosquito fly but prevents me, unaided, from flying? In spite of many thousands of years spent envying birds, we have only recently learned how to join them in the air. If people want to fly, we must build flying machines to carry us aloft.

Into the Wild Blue Yonder

Imagine taking a flight in a typical modern jetliner, such as a Boeing 737 or an Airbus A320. The airplane taxis from the terminal to the runway, where it pauses as if to gather strength (but actually to let another airliner get safely out of the way). Then the engines become very loud. The airliner accelerates down the runway for the better part of a mile, rises suddenly into the air, and climbs steeply. The landing gear retract with a series of thumps. Although few passengers notice, control surfaces on the wings (ailerons) and tail (elevators and a rudder) move during the flight to change or correct the craft's flight path. The craft climbs to its cruising altitude of 30,000 feet, so high that passengers would pass out from oxygen starvation if the cabin weren't pressurized. As the airplane nears its destination airport, the engines grow quieter. Elaborate, multipart surfaces (flaps) extend down and back from the rear of the wing, while a long narrow surface (a slat) along the front of each wing extends slightly forward. The airliner slows and descends. When the craft is a few minutes away from the airport, the landing gear extend with another series of thumps. As the airplane touches down, the engines are almost silent. When all the wheels are firmly on the ground, large flat plates (spoilers) on the top of the wing pop up, and the engines roar loudly as their thrust is reversed. Passengers are thrown forward against their seatbelts by the deceleration. After decelerating for a long way down the runway, the airplane turns off the runway and taxis to the terminal.

Now consider a bumblebee's typical flight. The bee climbs out of the burrow where she has built her nest. (A bee that is doing anything constructive around the nest is invariably a female; male drones literally live to mate and die shortly after doing their reproductive duty.) The morning is cool, so she vibrates her wings for several seconds to warm up her flight muscles, just as we shiver when we are cold. Once her muscles are warm enough, she launches into the air with a dainty hop and a vigorous wing beat. Her wings beat so fast that they are a faint blur to human eyes, flapping up and down about 150 times per second. She maneuvers—turning, rising, and falling—by adjusting how she beats her wings. If she flaps one side harder than the other, she will turn; if she flaps both sides harder, she will climb. If she adjusts her wing beat to tilt her body nose up, she will slow down. She will fly faster if she holds her body more horizontally. She flies fairly rapidly as she searches for flowers, making the occasional random turn if she does not find any. Eventually she sees the bright colors of a patch of flowers in a meadow. She flies straight at the nearest flower, gradually rearing up to slow down so that she is in a near-hover at the moment when she contacts the flower. She briskly goes about collecting pollen and packing it into "baskets" of bristles on her hind legs. The pollen will feed the queen's offspring when her eggs hatch back at the nest. She also takes a quick drink of nectar to refill her own fuel tank. When she finishes with the first flower, she steps off it directly into a hover and flies slowly to the next flower a few inches away. Again she pulls into a hover just before touchdown. She repeats this pattern, moving from flower to flower, until she has collected all the pollen she can carry. Launching directly into a climb from the last flower, she rises up a few meters above the ground and flies fairly directly back to her nest. Slowing and descending, she comes nearly to a hover just before she lands a few steps from the burrow entrance.

My descriptions of the airliner's and the bee's flights seem to have little in common except that both involve flight through the air, although this one similarity is clearly fundamental. The differences are striking. The airliner moves on the ground using retractable wheels; the bee uses legs that can fold up. The airliner needs a long takeoff and landing run on a dedicated runway; the bee casually hops into flight from wherever she happens to be standing. The airliner has fixed wings and a tail; the bee flaps her wings and is tailless. The airliner uses jet (gas turbine) engines that burn a kerosene-like fuel; the bee uses muscles that oxidize sugar and produce only a small amount of heat. The airliner's structure is made largely of metal; the bee's is a complex mixture of proteins and other organic chemicals. The one obvious structural similarity is that both possess wings.

Why Wings?

It is no coincidence that the airplane and the bee both have wings. Wings moving through the air provide a creature or a craft with a simple, effective way to produce lift (an upward force). Any flyer needs an upward force to counterbalance its weight, and by far the most economical way to produce lift is to use wings. A would-be flyer can use various brute-force tricks such as rocket engines or a crane to produce an upward force; but as we will discuss shortly, wings have the handy ability to convert a small push into a large lift. We take it for granted that birds and bugs use their wings to fly, and I suspect people have been taking it for granted since humans were first curious enough to notice such things. Knowing that wings are for flying, however, is not the same as knowing how wings work.

Until a century ago, people did not understand wings well enough to do anything useful with them. (Although venerable, useful items such as sails and windmill vanes actually function as wings, people did not recognize that fact until after the invention of airplanes.) We now know that a wing is a more or less planar structure with a large surface area that produces lift when it moves horizontally through the air. We also know quite a bit about how the wing produces that force.[1] In fact, wings are remarkable devices: in return for a relatively small horizontal push, they respond with a large upward force. A typical garden-variety wing—say, from a light airplane such as a Piper Cub— gives a lift force ten times greater than the horizontal push needed to move it through the air (as long as it is moving faster than some minimum speed). In other words, if I push on a wing with a force of 100 pounds to move it through the air, the wing produces 1,000 pounds of lift. Specialized high-performance wings, such as those used on competition sailplanes, can do much better, producing up to thirty-five or forty pounds of lift for every pound of horizontal push.

Wings take advantage of properties of air, so anything moving though air can potentially use wings. Effective wings, however, are not just perfectly flat plates. A simple flat plate can produce some lift, but it is not very effective and would not get either an animal or an airplane off the ground. An effective wing needs to have a particular set of shape properties, including a blunt front edge and a streamlined cross-section that tapers back to a thin, sharp trailing edge. Shape changes can have substantial effects on a wing's aerodynamic properties: lift, drag, minimum speed, and so on. Important shape traits include obvious ones such as overall area and more subtle ones such as the convexity of the upper and lower surfaces. These shape characteristics are what tailor a wing's performance to a particular mission. For instance, short broad wings (such as a sparrow's) are suited to fast maneuverable flight, whereas long narrow wings (such as a seagull's) are better for slower, less maneuverable, but more economical flight.[2] Of course, lift production is not the only constraint on a wing. Birds, for example, usually fold their wings compactly when not flying, which requires them to have joints and other structures not needed for lift production. Even so, the aerodynamic characteristics of a wing usually play the most important role. A fighter jet's short, broad, rigid wing is specialized for high speeds and violent maneuvering; but if such a wing were on an airliner or a single-engine Cessna, the craft would never get off the ground.

How Did They Get This Way? Conquering the Air

Flying animals and humans took very different routes to powered flight. Animal flight arose through the evolutionary process, whereas human flight came about through the directed efforts of a few talented people. Let us take a quick look at how flight arose in nature and technology.

The Natural Route

Biological evolution, acting largely though the process of natural selection, is an extremely slow, conservative process. Most animals and plants tend to be reasonably well adapted to their environments. Changes in an animal's anatomy or physiology, such as those caused by random genetic mutations, are almost always detrimental. That "almost" is crucial, however. Natural selection works on a timescale of hundreds of thousands or even millions of years. When a one-in-a-million beneficial change does occur, it tends to spread through the species and eventually become a normal feature.[3] Nature has produced wings several times during animal evolution. A variety of different animal groups, from fish to squirrels to lizards, have evolved wings for unpowered flight, or gliding. In contrast, powered flight requires such an intricate set of coordinated properties that the right combination has only occurred four times in the 450-million-year history of land animals. The four fortunate groups that have evolved powered flight are insects, pterosaurs, birds, and bats.

The details of how flight arose in animals have long been a matter of debate among paleontologists and biologists. Our knowledge of ancient animals is incomplete (not all times and places have produced fossils), and the early stages of the evolution of flight occurred during gaps in our current fossil record. Although frustrated by this missing information, scientists are not surprised by it. Evolution tends to occur in fits and starts: innovations appear quickly and spread rapidly but are followed by long periods of stasis, or lack of change.[4] Each time flight arose, it probably spread rapidly (in geological terms, that means a few tens of thousands of years) because of its great benefits. As far as we can tell, it did so at times and places that left no fossils. Fossils form best under water, and animals always seem to have begun the evolution of flight far from water.

Insects were the first flyers, evolving wings and powered flight more than 400 million years ago. Pterosaurs, extinct relatives of dinosaurs, were the next animals to achieve flight (about 200 million years ago), followed by birds 40 or 50 million years later. Bats were the most recent group to evolve flight, becoming aerial at least 55 million years ago.[5] These dates represent our best estimates based on the oldest known fossils in each group; any of the groups could well have evolved much earlier.

Among scientists who study the origins of animal flight, the largest number have argued most heatedly about the origin of flight in birds. At first glance, this seems odd, given the existence of the fabulous *Archaeopteryx* fossils. Several specimens were found in the mid-1800s (and a few more since then), and they are so complete that they preserve details of feathers and other soft body parts. *Archaeopteryx* seems as if it should be the perfect missing link, or transition fossil. The animal had characteristics of both birds (particularly wings and feathers) and dinosaurs (teeth, scaly skin, and a long bony tail). For more than a century, scientists based arguments about the evolution of flight on evidence from *Archaeopteryx*.[6] Nevertheless, its wings are so similar in shape to those of some modern birds that *Archaeopteryx* actually doesn't tell us much more about the evolution of flight than we could learn from the birds around us. While it does offer tantalizing tidbits about its ancestors, it unfortunately tells us little about the evolutionary process that led to flight.

The debate about the origin of bird flight has become heated over the years and centers around two opposing viewpoints. Some scientists argue that flight evolved from the ground up: running dinosaurs leaping for flying insect prey gradually accumulated wing-like adaptations that allowed them to jump higher and farther, and they also evolved flapping so that they could run faster and prolong their leaps. The eminent Yale paleontologist John Ostrom has long been a vocal champion of this view, which many paleontologists have supported over the years.[7] Other scientists argue that bird flight began from the trees down. They imagine a small, insect-eating, tree-dwelling reptile living in an environment in which a squirrel would be comfortable. Some of these reptiles, like some squirrels, evolved wings to glide from tree to tree. Unlike squirrels, however, the surfaces of their wings were made of feathers rather than flaps of stretched skin. Wing movements for steering evolved into flapping, at first just to prolong glides. Eventually, flapping allowed for true powered flight. The widely published ornithologist Alan Feduccia has argued for the tree-down scenario, with support from a number of biomechanics researchers, such as Jeremy Rayner from the United Kingdom and Ulla Norberg from Sweden.[8]

For more than a century, the pendulum has swung back and forth between the ground-up and the trees-down theories. Just as one seems to be gaining momentum, scientists find new evidence or develop new suggestions to support the other. They have also made a few attempts over the years to reconcile or merge the two scenarios, with little lasting success. A recent suggestion by British researchers Joseph Garner, Graham Taylor, and Adrian Thomas, dubbed the "pouncing proavis" model, borrows elements from both theories and combines them in a way that agrees with the newest fossil evidence: that non-bird dinosaurs had what appear to be feathers.[9] In this model, a leaping ambush predator evolved feathered hand- or arm-mounted control surfaces to better direct leaps at prey. The animal probably started as a runner; but as its control surfaces enlarged and became more wing-like, its pouncing attacks became short glides (swoops), and the animal launched attacks from elevated perches (such as rocks and tree branches). Short glides became long glides and control movements became flapping until true powered flight evolved. As plausible (and well supported by the evidence) as the pouncing proavis model is, the ground-up and trees-down camps today seem almost as far apart as ever. Garner and his colleagues published their suggestion in 1999; but so far, the major supporters of the traditional models have hardly acknowledged this new idea.

The evolution of flight in bats has received much less scientific attention, probably because no one has yet found the equivalent of a captivating transition fossil such as *Archaeopteryx*. In contrast to the bird case, however, scientists are in broad general agreement that bats evolved flight from the trees down, based on the fact that their nearest relatives are all arboreal as well as on some details of their anatomy. For bats, the scenario goes like this: bat ancestors lived in trees and ate the insects they found there. These ancestors leaped among tree branches and would have benefited from any modifications that allowed them to extend or steer their leaps. Front limbs elongated, and flaps of skin developed between fingers and from limb to body. Once the front limbs became long enough and the stretched skin had enough surface area, the animals became effective gliders, something like today's flying squirrels. Steering movements evolved into flapping, first to lengthen glides and then for true powered flight. Once they had evolved powered flight, the animals would have been essentially indistinguishable from modern bats.[10]

An analogous process would have occurred in the other two groups that evolved powered flight: pterosaurs and insects. Yet as far as we can tell, flight evolved only once in each of these groups. Nature must require a rather unusual set of circumstances for the production of powered flight if it has only evolved four times in the hundreds of millions of years of animal history. And considering the requirements of flight, we should not be surprised. An animal needs a combination of anatomical attributes (large surfaces that can be co-opted for an aerodynamic function) and nervous system sophistication that allows for at least rudimentary flight control, along with environmental or ecological conditions in which limited flight ability are of great benefit. Clearly, this combination, along with the necessary mutations to take advantage of it, is quite rare.

The Artificial Route

Airplanes developed as a result of guided technological progress. Naïvely, we might say that airplanes evolved many times because many researchers and inventors built flying machines immediately before and after the Wright brothers flew. With the exception of a small handful of gliders, however, none of these flying machines were successful before the Wrights', and few were even marginally successful for many years after the Wrights first flew.[11] In my opinion, human flight originated just once, and Orville and Wilbur Wright were the true inventors of the airplane. All successful airplanes, even today, are designed using the brothers' fundamental insight about the proper way to control a flying machine. Those machines that hopped briefly into the air before the Wrights flew were essentially uncontrollable and thus were dead ends for further development. After the Wrights had flown, several inventors built machines that could get off the ground; but none were truly flyable until the Wrights publicly demonstrated the proper way to turn an airplane.[12] Within a year or two of Wilbur Wright's 1908 European tour, French and British inventors were building airplanes that would actually go where the pilot wanted them to go rather than in barely controlled, more or less straight lines. At that point, just before World War I, the initiative passed from the Wrights to their successors, and most progress in aeronautics occurred in Europe for the next decade or so.[13]

What is this process of guided technological progress? Clearly, it is enormously faster than biological evolution. Changes that might take hundreds of thousands of years of animal evolution can take place in less than a decade of technological development. The key is that people can guide the process of technological development toward some goal rather than wait for random changes to produce the right combination of building blocks. Technological progress requires people with vision, who can foresee the goal, as well as people with the practical knowledge needed to work toward the goal. If one person possesses both traits, so much the better. Finally, the background level of technology must be advanced enough so that the designer can make use of the necessary building blocks without inventing deeper layers. In other words, progress can occur only when current technology provides the necessary tools. If the designer must also build the tools to make the required tools, contemporary technology is probably not up to achieving the designer's goal. Despite their engineering prowess, the ancient Romans could not have developed powered flight. They had neither the materials (machine tools, interchangeable parts, steel wire, aluminum alloys) nor the knowledge base (basic air properties, the experimental approach) to tackle the task. Similarly, in 1900, when engineers were just figuring out how to build low-powered gasoline engines and cool them enough to keep them from melting down, the appropriate tools and materials to build a jet engine simply did not exist, even if an engineer could have correctly designed one.

The Wright brothers succeeded because they had ideal qualities for the task: they possessed the vision to see the goal—powered flight— along with tremendous practical knowledge of machinery. Often described as "bicycle mechanics," the two men were much more than that. They first gained practical knowledge of complex machinery by working as printers. After designing and building his first printing press, Orville worked for two years as an apprentice printer. Later, he and Wilbur designed and built several increasingly sophisticated printing presses for themselves and other local printing shops.[14] They ran their own printing shop and published their own newspaper. They also got in on the ground floor of the turn-of-the-century bicycle craze. First, they repaired and sold mass-produced bicycles; then they produced their own line of high-quality bicycles.[15] They were thus intimately familiar with designing and building a variety of kinds of complex machinery. Intellectually, they seemed to possess a sort of synergistic genius. They debated and argued over problems, playing off each other as devil's advocates and hashing out possible solutions that neither alone might have imagined. Each was a highly intelligent man, but their collaboration produced true genius.

Although they had no formal scientific training, Orville and Wilbur were widely read and brilliant intuitive researchers. They gathered and read all the relevant technical literature that they could find and proceeded through a careful step-by-step set of experiments to solve different aspects of the flying problem. First, they built a large kite to test their control system. Once they were confident that their system worked, they built a man-carrying glider to confirm their control arrangement and aerodynamic calculations and, not incidentally, to teach themselves to fly. When their first glider did not live up to expectations, they went back to basics. They built a wind tunnel and sophisticated and clever instruments to measure the aerodynamic properties of various wing models. They used these measurements to design their next glider.

The second glider lived up to their calculations, and they learned a great deal about flying in that craft. Still, it had turning problems. They analyzed the problems, added a rudder to their next glider, modified the control system, and proceeded to log hundreds of successful flights. This third glider was the craft in which they really learned to fly, and it gave them confidence to proceed to the next step.[16] Now all they needed was a light, powerful engine and propellers to spin.

When they could not locate an appropriate engine (and after being told it would cost a fortune and take forever to have one custom-built), they designed and built their own. (Actually, Orville did most of the designing, while their machinist, Charles Taylor, did most of the building.)[17] It was crude even by the standards of the day—for example, it substituted exhaust heat for a carburetor to vaporize fuel—but it was relatively light, produced more power than the minimum they needed, and the price was right. At about the same time, the Wrights also discovered that essentially nothing scientific was known about propellers (even though screw propellers had been in use on ships for nearly a century). They were apparently the first people to realize that propellers are just rotating wings, and they applied their wing research to designing efficient propellers. No one else built better ones for more than a decade.[18] They scaled up the successful glider, mounted the engine, and, on December 17, 1903, made history.

The Wright brothers and their achievements illustrate a number of key qualities needed for technological progress. Their combination of vision, practical knowledge, innate understanding of research, creativity, and intelligence made them the right people for the task. Also, the time was right. The necessary technological underpinnings—gasoline engines, power-driven tools for working wood and metal, Newtonian physics, and the beginnings of fluid mechanics—were present and handy. Failures, rather than acting as dead ends, spurred the brothers to take a different route or to back up and review or replace underlying assumptions. When the glider designed according to Otto Lilienthal's tables did not produce as much lift as they calculated that it should, they went back to the beginning and measured some of those basic numbers for themselves. In a nutshell, technological progress can be millions of times faster than biological evolution because people direct it toward a goal; and they can observe, understand, and attempt to correct errors.

Does Mother Nature Know Best?
Critters versus Craft

Today many people tend to equate *natural* with *good* and *artificial* or *technological* with *bad*. If the popularity of the phrase "all natural" on the labels of supermarket items is any indication, advertisers certainly believe this attitude is common. According to this way of thinking, birds are good at flying, while airplanes are, if not bad, at least not as good. Engineers, in contrast, tend to view biological structures with suspicion and machinery with satisfaction. An engineer understands the function and the underlying design principles of a Boeing 747, whereas he or she has at best a weak grasp of the mission of a fly or a crow, let alone the design constraints imposed by nature. To an engineer, an airplane can be a work of art whereas a robin is a fragile, wet bundle of exotic materials with an unfathomable control system (brain) and a mysterious purpose.

Both views are shortsighted. Flying animals tend to be well suited to do what they need to do, and airplanes (successful ones, at least) are well suited to their requirements. Flying animals are highly integrated, specialized flyers: their nervous systems are superbly adapted to operate their flight systems in a direct and intimate fashion that human pilots can only envy. Airplanes and pilots will never be as well integrated, although computerized, automatically stabilizing control systems in the most advanced airplanes show striking parallels to reflex stabilizing responses in flying animals. No airplane will ever be as agile or versatile as the clumsiest flying animal nor be able to repair or replace itself as all animals do routinely. On technology's plus side, airplane production is not tied to the human reproductive cycle; designers produce many generations of airplane improvements over the span of one human generation. Moreover, no animal will ever fly as fast as most airplanes can or as high as high-flying airplanes do, and none will ever carry a load as big as the payload that even small airplanes can carry. Animals operate on a much smaller, slower scale than airplanes do.

In addition to their obvious size differences, flying animals and airplanes are different because their missions are different. An airliner would be very bad at landing on flowers and collecting pollen, just as a bee would be equally bad at carrying large groups of people from one place to another. The bee has an appropriate design for collecting pollen and digging nests, and the airliner is suitable for carrying large loads for long distances at high speeds. Although not perfect, each must do reasonably well at what it does because, in both the biological and the commercial realm, if you are not good enough, something will replace you that is. All in all, flying animals and airplanes are different. Neither flies better than the other in any general sense, but they are reasonably well suited to their separate tasks.

The Issue of Design

People deliberately design airplanes, so no one worries about the question of whether or not airplanes are designed. Can we describe flying animals as having a design? From a scientific perspective, we have no evidence of a designer (questions of faith and deities are impossible to measure objectively so are outside the arena of science). Any casual observer, however, can easily see that different animals are adapted to accomplish different missions. For example, bees and hummingbirds are clearly adapted to visit flowers, most bats are adapted to catch and eat insects in flight, vultures are adapted to finding and eating carrion, and pelicans are adapted to fishing and a generally amphibious lifestyle. These animals have been fitted by evolution to carry out appropriate tasks reasonably well and better than their competitors. Speaking of the design of an animal makes sense because aspects of its body form and function are well suited to particular ways of making a living. Pelicans' expandable bills make effective fish collection devices, and their webbed feet provide locomotion on both land and water. Thus, the pelican design includes features for an oceanic, fish-eating lifestyle. Likewise, the vulture design includes efficient soaring flight to search for carrion over large areas, good vision to spot corpses at great distances, and a sharp, heavy beak to tear through fur and skin. In this sense, the notion of design is perfectly valid and does not require a designer beyond the action provided by natural selection. This, then, is what I mean throughout the book when I refer to an animal's design.

Where to from Here?

In the chapters that follow, I will compare animal and airplane flight by looking at the tasks that any flyer must perform, whether it is a gnat or a jumbo jet, and then at the techniques that have evolved or been engineered to perform said tasks. The latter will be much like following a railroad route that is usually two separate, parallel tracks but occasionally merges into a single track. In most cases, animals and airplanes use different techniques to carry out a given task (flapping wings versus propellers, for example), but sometimes they do the same job with essentially the same means (such as performing turns). I will conclude with a look at why animal and airplane designs have followed different routes and why effective airplane and animal designs require these different approaches.

~ 2 ~

Hey, Buddy, Need a Lift?

We are so used to seeing birds and airplanes that we take wings for granted. This is a shame because wings are remarkable. A jumbo jet weighs three or four times more than a diesel railroad locomotive, yet the airliner's wings keep it aloft using nothing but thin air. Geese use their wings to fly over the Himalayas, and hummingbirds use theirs to fly backward and sideways.

Despite their different uses, all wings obey the same basic aerodynamic principles. One of the most important is that, simply for the price of movement through the air, a wing rewards a flyer with a substantial upward force that can overcome the flyer's weight and allow it to fly. As a result, wings are by far the most practical, efficient way to fly, which is why all flying animals and all successful flying machines use them.

When it moves through the air, a properly shaped wing produces a force at right angles to its movement. If the wing moves horizontally, this force is directed up and becomes lift. Weight is also a force: a downward force caused by the gravitational attraction of the earth. When a person steps on a bathroom scale, weight is how hard he or she pushes down on the scale. A wing can fly if it can produce enough force to overcome its own weight plus the weight (or aerodynamic loads) of anything else attached to it. I suspect that most people believe

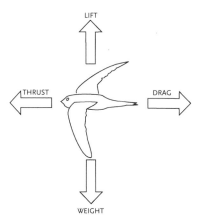

Figure 2.1 The four forces experienced by any flying animal or aircraft.

that a wing needs to produce more lift than the weight it is carrying in order to stay aloft, but that is not necessarily true. The wing does need to produce more lift than weight to initiate a climb; but in a steady climb, as in level flight, the lift on an airplane's wing exactly balances the weight. Thus, lift and weight are equal and opposite (fig. 2.1).

The difference between our everyday use of the word *lift* and its aerodynamic definition can be confusing. In horizontal flight, lift is reassuringly straight up; but if the wing descends (gliding downward), the lift tilts forward. If it ascends, the lift tilts back. For flying animals, this change in the angle of lift is crucial. As an animal flaps its wings, the lift at any instant may point up, up and forward, mostly forward, or even backward, depending on how the animal moves its wings at any point in the wing stroke. In fact, the whole point of flapping is to allow the wings to produce lift that is, on average, tilted forward and thus pulls the animal through the air.[1]

When a wing stands still, it does not do anything special; but when it moves through the air, it produces lift due to its shape and orientation. Moving a wing through the air comes with both a cost and an opportunity. Air resists the wing's movement and pushes back with a resisting force called drag. All flyers produce a forward force called thrust to overcome drag, meaning that thrust production is the cost of flight (fig. 2.1). A well-designed wing generates much more lift than drag. In other words, in response to a relatively small forward push, a wing produces a much larger upward pull. Thus, the opportunity comes from how lift increases with speed. Intuitively, we might expect a wing to produce more lift as it goes faster; but in fact, a wing produces much more lift in return for going only a little faster. The wing's lift is proportional to the square of its speed, so a modest increase in speed gives a big increase in lift.[2] For example, all else being equal, a wing that produces one hundred pounds of lift at twenty miles per hour will produce four hundred pounds of lift at forty miles per hour.

How much more lift does a wing make than the thrust needed to overcome its drag? Usually, a lot more. In fact, the maximum ratio of lift to drag is one of the most important traits of a wing, and engineers consider it to be the prime measure of a wing's efficiency.[3] (A wing's lift-to-drag ratio actually changes depending on its orientation, but generally we are interested in the lift-to-drag ratio in the orientation that gives the highest value.) A typical wing—say, from a large bird or a small airplane—will have a lift-to-drag ratio of about ten to one. In other words, it generates ten times as much lift as the thrust needed to overcome its drag. If I push on this wing with a force of twenty pounds, I get two hundred pounds of lift. This is quite a bargain and is what makes flight practical. Without wings, I have to come up with some way to produce the entire force to offset my weight, perhaps using something like a rocket engine. Rockets do fly, but they are nowhere near as practical or efficient as airplanes.

Various flyers have a broad range of lift-to-drag ratios. The wings of the humble fruit fly buzzing around your overripe banana have a lift-to-drag ratio of just under two to one, mainly a consequence of the insect's small size and stubby wings. Bumblebees do better at about three to one, starlings at five to one, and crows at about six to one.[4] Long-winged animals do better yet, with swifts at ten to one and gulls at eleven to one.[5] Animals that soar for a living (discussed in Chapter 7) need high lift-to-drag ratios, so hawks and vultures range from ten to one up to about sixteen to one, and wandering albatrosses are the hands-down champs of the animal world at nineteen to one.[6] Size also favors airplanes. Small, light, four-seat, single-engine airplanes such as the Cessna 172 are usually around ten to one, with large airliners doing better at fifteen to one. Sailplanes (specialized soaring airplanes without engines) do best of all, with lift-to-drag ratios starting at thirty to one and sometimes going as high as sixty to one.[7]

Making Wings More Effective: More Lift, Less Drag

If I hold my hand out the window of a car that is driving down a high-way, I can make it act like a wing. When I hold my hand flat or slightly cupped, palm down, I clearly feel how it pulls up when I tilt up the front edge. My hand is a fairly miserable wing, however. It probably has a lift-to-drag ratio of little better than one to one and barely produces enough lift to support its own weight, let alone my arm or my whole body. For a wing to be truly effective and useful, it needs to have the properties that I call the five features of highly effective wings.

Surface Area

The most obvious feature of any useful wing is its surface area. A wing's surface area is usually measured as its planform area, or the area you would see if you were looking down on the wing from directly above. The amount of lift a wing produces is proportional to its surface area.[8] Not all parts of the wing contribute equally to the total lift because a wing usually generates the greatest upward pull near the root and less and less toward the tip. Even so, increasing a wing's area increases its lift because each square inch of wing surface contributes to the total lift: the more square inches, the more lift.

Aspect Ratio

The aspect ratio is a wing's length from tip to tip (its span) divided by its average width from front to back, or its mean chord. Long, narrow wings have high aspect ratios; short, stubby wings have low aspect ratios. A wing's aspect ratio is important because it strongly influences the lift-to-drag ratio. Long, narrow wings, which have high aspect ratios, have higher lift-to-drag ratios than do short, broad wings of the same area, which have low aspect ratios. High-aspect-ratio wings pay less of a drag penalty to produce a given amount of lift, making them more effective.

A wing's aspect ratio affects its lift-to-drag ratio because the processes that produce lift also produce an additional source of drag: induced drag. The physical mechanism that produces extra drag is complex, but imagine it as a function of wingtip size in relation to the wing's area. A long, narrow wing has a small tip for its area and so has less drag. A short, broad, low-aspect-ratio wing has a lot of tip for its area and so suffers more drag. If both wings have similar surface area, both will produce about the same amount of lift; but the narrow, high-aspect-ratio wing will produce less drag, so its lift-to-drag ratio will be higher.[9]

Useful wings for subsonic speeds don't generally have aspect ratios much lower than five, and the wings of the Wright brothers' first airplane had a fairly typical aspect ratio of six.[10] Small birds have aspect ratios ranging from about five for a sparrow to about eleven for a swift. Larger birds cover a similar range, from about five for a crow to about ten for a heron. Jet airliners typically have aspect ratios of about seven or eight. Only the specialized soarers (discussed in Chapter 7) deviate much from this range, with albatrosses coming in at about fifteen and sailplanes to a range between twenty and thirty.[11] A long, slender wing with an aspect ratio of thirty has about the same aspect ratio as a one-inch-wide Venetian mini-blind slat in a thirty-inch-wide window. That's one skinny wing!

As I have mentioned, animals and airplanes with high-aspect-ratio wings also have high lift-to-drag ratios. If lift-to-drag ratio is a measure of a wing's effectiveness, why don't all wings have high aspect ratios? The answer illustrates a classic design trade-off: longer, narrower wings are more efficient aerodynamically, but they also tend to be heavier or more fragile than shorter wings. When a wing gets too long and skinny, the only way to make it strong enough to carry flight loads is to strengthen it to a point of unacceptable heaviness.

All wings face functional trade-offs as well. Effective wings tend to be large and cumbersome on the ground: the longer a wing, the more awkward and unwieldy it is in tight quarters. Therefore, birds that fly through forests, such as sparrows and woodpeckers, tend to have low-aspect-ratio wings. Similarly, the space available at airport gates places real limits on the maximum wingspan of airliners. This limit became a serious concern when Airbus was designing its new A380 super jumbo airliner. Designers had to settle on slightly shorter, broader wings than they preferred due to airport gate restrictions. Even though airliners might gain some fuel efficiency with longer, narrower wings, they would no longer fit into the gates at most airports.

Modifying a wing's tip can make the wing perform as if it had a higher aspect ratio. For example, many birds fly with the wingtip's primary feathers fanned out, leaving a little space between each feather as you might spread the fingers of your hand. Biologists measure higher lift-to-drag ratios on such wings compared to those without separated primary feathers.[12] At first, engineers tried to imitate separated primaries with various wingtip extensions and slots, but they eventually developed a simpler structure, usually called a winglet. Winglets, which are little vertical or near-vertical fins on the wingtip, look like small extensions of the wing that have been bent upward. Designing winglets that actually improve a wing's performance is challenging because slight changes in a winglet's size, shape, and angle can have significant effects on its performance. In fact, I have been told by knowledgeable aeronautical engineers that quite a few airplanes have winglets designed for looks rather than for aerodynamic effect. A well-designed winglet can, however, improve the wing's lift-to-drag ratio enough to shave a percentage point or two off fuel consumption, no small thing when a jumbo jet burns more than 25,000 gallons of fuel on one transatlantic flight.

Orientation

For a wing to make lift, it must deflect air downward. So as it moves through the air, a wing pushes air down, and the air pushes back with an equal and opposite reaction. When I hold my up-tilted hand out of the window of a moving car, I feel the air pushing my hand up. Tilting my hand up makes it deflect more air downward, which pushes my hand up harder. Wings work in generally the same way, although a decent wing produces much more lift at a much smaller angle. Airplane designers are intensely interested in how much lift a wing produces under different conditions, one of which is called the angle of attack—the angle between the wing and the direction it moves through the air. Imagine a rod running from the front of a wing (the leading edge) to the back of the wing (the trailing edge) (fig. 2.2).

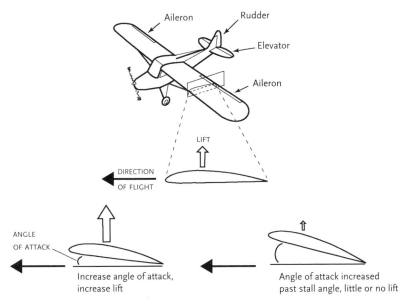

Figure 2.2 Airplane control surfaces and airfoil (wing cross-section). As the airfoil's angle of attack changes, lift changes: increasing the angle of attack increases lift until the stall angle is reached. At the stall angle, lift decreases dramatically.

If the wing moves horizontally through the air and the rod is also horizontal, the wing will produce little or no lift. Tilt the wing up slightly (say, by five degrees) and move it horizontally, and the wing produces much more lift. Tilt it up even more (to ten degrees), and the wing produces approximately twice as much lift as it did at five degrees. So increasing a wing's angle of attack increases its lift.[13]

An effective wing thus needs some mechanism for adjusting its angle of attack. Conventional airplanes make this adjustment by using horizontal tail control surfaces called elevators, whereas flying animals mostly do it by swinging their wings forward or backward, shifting their center of gravity relative to their wings. Hang gliders also use weight shifting to control their angle of attack.

Obviously, increasing the angle of attack to increase lift must have limits, or 747s and crows and houseflies would all be flying around with their wings oriented vertically instead of horizontally. The limiting factor is a process called stall, which has nothing whatsoever to do with engines. When a wing stalls, the air no longer flows smoothly over its top surface. Instead, smooth air flow moves away from the surface, leaving a large turbulent zone on top of the wing. When this happens, the wing's lift drops dramatically, and simultaneously its drag rises sharply. When an airplane's or a bird's wing stalls, the flyer stops flying and starts falling, meaning that stalling is not good in most situations. The angle at which a wing stalls, known as the critical angle of attack, varies with wing design and size. Typical airplanes stall at angles ranging from twelve to twenty degrees, birds usually stall between fifteen and thirty-five degrees, and insect wings don't stall until they reach angles of twenty-five to fifty degrees.[14] In most cases, the bigger the flyer, the lower the critical angle.

Structural Strength

Wings need to be strong enough to carry the flyer's aerodynamic loads. Obviously, that means they need to be strong enough to at least carry an animal's or an airplane's maximum weight. In practice, however, they need to be even stronger. In a relatively gentle turn, centrifugal force can increase weight on the wings by 50 percent, and a sharp turn might triple or even quadruple the weight. When a fighter pilot describes a "6 G turn," he or she means that the turn is so sharp that centrifugal force causes the pilot's weight *and* the weight of the airplane to become six times heavier than normal, forcing the wing to carry six times the airplane's normal weight. The airplanes in the Federal Aviation Administration's utility category (craft not designed for aerobatics) typically have wings designed to withstand a four-and-a-half-fold increase in weight, while specialized aerobatic airplanes are designed to handle six-fold weight increases.[15] Combat aircraft are often designed to handle even higher weight increases.

Why not design all wings to handle ten- or fifteen-fold weight increases so as not to worry about overstressing any of them? Unfortunately, a stronger wing is also heavier, and extra weight is something that all flyers strive to avoid. Birds have air-filled bones, and modern airliners use expensive materials such as titanium and carbon-fiber composites to maintain strength and keep down weight.

Once again, we encounter a trade-off: strength versus lightness. A heavier structure requires a flyer to fly faster, which produces more drag, which puts more stress on the wing and increases fuel consumption. So wings need to be as light as possible while still being strong enough to do the job.

The logical approach is to design a wing to be just strong enough to handle the highest load it would typically encounter and then to add a certain amount of strength as a safety net for unexpected events. The key here is to figure out how much extra strength is necessary. A conservative engineer might want a safety factor of two or more (making the structure capable of handling at least twice the expected maximum load), but a factor that large might lead to unacceptably heavy wings. For most modern airplanes, the Federal Aviation Administration mandates a minimum safety factor of one and a half.[16] The limb bones of a wide range of running animals have safety factors ranging from about 1.6 to 4.8, which more or less matches the safety factor of wing bones in the few birds that have been studied so far.[17] Given that airplanes do fine with safety factors of 1.5, why do some animals appear to have much higher ones? The answer is that researchers may be underestimating peak natural loads, or there is a wide range of natural variation among animals (or both).

Considering their strength, wings are some of the lightest structures in the animal kingdom. Insects, for example, have an ingenious arrangement of rod-like veins supporting extremely thin sheets of flexible membrane. The main veins radiate from the wing base to the wingtip and trailing edge. Secondary veins run between the main veins to connect and reinforce them (fig. 2.3). The veins are not all in a flat plane; they alternate, with some slightly higher or lower than others, giving the cross-section of an insect wing a somewhat pleated appearance. This arrangement helps the wing resist lengthwise bending (when the tip is bent up or down) while allowing it to twist and bend front to back. The ability to resist lengthwise bending means that the insect is better able to carry flight loads, whereas twisting and bending the trailing edge up and down improves the effectiveness of flapping. These wings use very little material (the wing membrane

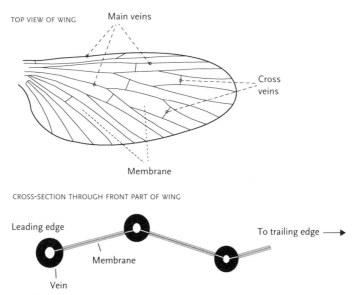

Figure 2.3 Structure of a typical insect wing.

may be less than 1/10,000-inch thick), so they are extremely light for their strength.

The earliest airplanes, from those of the Wright brothers through craft constructed during the 1920s, were built of wood and covered with fabric that was painted with a lacquer-like sealant called dope.[18] Wood and fabric are obviously lighter than metal, and early aircraft engines had too little power to lift anything heavier. Most of these airplanes were biplanes because the two sets of wings can easily be strengthened with bracing wires running from wing to wing. Biplane wings and bracing wires add drag; but at the low speeds of the early airplanes, designers accepted the small additional drag in preference to heavier wings.

As engines became more powerful and flight speeds increased, wood became too fragile to carry aerodynamic loads. Engineers also realized that the drag of external bracing wires cost them too much lost speed and range. First, they replaced wooden frames with tubular steel frames covered with fabric or wood. Then, as high-strength aluminum alloys became available, the typical airplane configuration changed. It became a monoplane with cantilever wings (wings with no external bracing) and an aluminum-skinned, semi-monocoque structure (the skin itself carries some or most of the structural load) with a reduced internal framework. This was the typical arrangement for large or high-performance airplanes from the 1930s into the 1970s and is still common today among small and medium-sized airplanes.[19]

In the 1970s and 1980s, materials scientists developed lightweight, high-strength composites. A composite is made from two or more separate materials that have very different properties. Fiberglass is a composite that has been around for decades. It consists of a cloth-like material of fine glass threads embedded in a type of plastic called polyester resin. Cured resin is a stiff, brittle material, while glass cloth is a floppy fabric. Together, however, the resin gives the glass cloth a fixed shape, and the cloth overcomes the resin's brittleness and gives it strength and toughness.

Although several homebuilt airplane designs use fiberglass skins, most airplane designers consider fiberglass to be too heavy for airplane use. New fibers, lighter and stronger than glass, are used in aerospace structures. These include boron, graphite (also known as carbon fiber), and Kevlar, usually embedded in a high-tech, lightweight, heat-setting epoxy resin.[20] For their weight, these composites are considerably stronger than steel or aluminum alloy. Working with composites, however, is very different from working with metals. Construction techniques are in many ways more akin to techniques of the textile industry. A major advantage of this approach is that composite parts can be very precisely tailored to their expected loads. Builders can add another layer of graphite fabric to add strength or adjust the orientation of the fibers in the fabric to improve the distribution of structural loads.[21]

Shape: Trailing Edge and Camber

If you could slice a typical wing from front to back and look at the cut edge, you would see what aeronautical engineers call the wing section, or airfoil. The cross-section would reveal two important aspects of wing shape. First, the trailing edge comes to a sharp point rather than a blunt edge, a feature that greatly improves lift production. A sharp trailing edge allows air from both the top and the bottom to flow smoothly off the back of the wing, which is crucial for the airflow pattern that generates lift.[22] Second, on most wings, the top is a bit more convex than the bottom surface, making the wing look slightly humped up.

This difference in convexity is called the camber; and as figure 2.4 illustrates, the camber was dramatic on the 1903 Wright Flyer and is also very pronounced on typical bird wings. Camber is usually more modest on modern airplane wings, mainly because engineers have discovered that a concave or hollowed-out bottom surface does not yield enough aerodynamic improvement to justify the structural weakness of thin wings. Camber is measured by how much a line following the middle of the airfoil deviates from a straight line connecting the front and back of the airfoil, usually expressed as a percentage

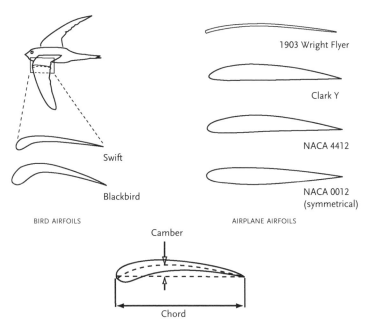

Figure 2.4 A variety of cambered and uncambered airfoils. Camber is measured as maximum distance of the airfoil midline from the chord divided by the chord length.

of the length of the straight line (or the chord). Camber can increase the amount of lift that a wing produces at any given angle of attack without much increase in drag.[23] To understand why this happens, consider the trailing edge of a wing with a symmetrical airfoil—that is, with no camber. Air flowing over this wing tends to move straight back with no deflection. Now consider a cambered airfoil. The camber effectively bends down the trailing edge; so even at an angle of attack of zero, air flowing over the airfoil streams off the trailing edge with a slight downward deflection. Therefore, at any given angle of attack, a cambered wing tends to have more lift but about the same amount of drag as a wing with no camber.

Camber is beneficial only within a fairly narrow range. If the wing is too humped up, drag starts to rise faster than lift; and a wing with exaggerated camber will have excessive drag and will stall at low angles of attack. Because flying animals are both smaller and slower than most airplanes, their wings tend to be less sensitive to the increased drag penalty of high camber. Birds, for example, may have camber as high as 14 or 15 percent, whereas airplane airfoils rarely go as high as 14 percent; 5 or 6 percent is more typical.[24]

Animals generally use a good deal of camber on the downstroke and reduce or eliminate it on the upstroke. Most birds cannot entirely eliminate camber because a certain amount is built into the shape of their feathers. Bats and insects, in contrast, have wings that are cleverly arranged to let them radically change the camber. Bats support their wings with enormously elongated finger bones, like the supports of an umbrella, and they can flatten or curve a wing as easily as I can make a fist. Insects have some ability to adjust the veins in a wing, so they can also hump up or flatten it out as needed.

How much camber should a wing have? The shape of the airfoil affects traits such as stall angle, maximum lift, maximum lift-to-drag ratio, and wing thickness. High camber is useful in some conditions, particularly at lower speeds. At higher speeds, less camber works better, which is why airplane wings generally have less camber than birds do. In fact, more than a few airplanes fly with zero-camber airfoils, usually called symmetrical airfoils. Many supersonic airplanes have symmetrical airfoils, as did at least one Second World War fighter (the Bell P-39 Airacobra) and one small jet airliner (the Fokker F-28 Fellowship). A wing with a symmetrical airfoil can still produce lift; it just does so entirely with angle of attack. Airplanes that fly at high speed can often get by without any camber because they get plenty of lift with a modest angle of attack. Moreover, at high speeds, cambered wings may suffer more significant drag penalties than they do at lower speeds. Aerobatic airplanes that need to be able to fly equally well when they are upside down and right side up (such as the Pitts Special) also use zero-camber wings. These airplanes often compensate for slightly less efficient wings with significantly more powerful engines.

Feathered or Riveted: Wings of All Kinds

Regardless of whether a wing is attached to a mosquito or a jet airliner, if we want to increase its effectiveness (its lift-to-drag ratio), we must increase its surface area, its aspect ratio, its angle of attack, or its camber, or some combination of these elements. We also must work to create the lightest possible structure that is strong enough to carry the load. Even though maximizing a wing's effectiveness is beneficial for all wings, aerodynamic efficiency is never the only constraint. Other constraints, such as how fast a wing flies or what it is made of, account for the huge variety of animal and airplane wings and their enormous range of sizes, shapes, and structures.

Size

The size of a wing can have a significant effect on its lift-to-drag ratio. Scientists and engineers use a dimensionless index called the Reynolds number to quantify size effects in air and water. The Reynolds number (Re) is defined by the equation

$$Re = (\text{air density} \times \text{air speed} \times \text{length})/\text{viscosity}.$$

Technically, the Reynolds number is the ratio of drag forces caused by pressure difference to drag forces caused by viscosity—the stickiness or thickness of a liquid or gas.[25] In practical terms, small, slow flyers such as insects or small birds fly at low Reynolds numbers

(500 to 10,000), whereas airplanes fly at Reynolds numbers of 1 million or higher. When wings operate at low Reynolds numbers, they have thicker boundary layers, which are zones close to an object's surface where friction slows the flow of air. Thicker boundary layers mean that a wing will produce less lift and more drag.[26] So the smallest flyers have relatively inefficient wings compared to those of large airplanes.

A house fly's or a grasshopper's wings produce much less lift per square inch of surface area than do a pigeon's, and a pigeon's wings produce much less lift per square inch than do a Piper Cub's wings. Insects, the smallest flyers, compensate in a couple of different ways. First, they have extremely light, strong wings for their size. (Insects can get away with their pleated wing structure because the boundary layers around their wings are so thick at their low Reynolds numbers.) Small animals also have disproportionately strong muscles (see Chapter 3), so they can afford to apply more muscle power to their wings than larger flyers can. Some insects even make a virtue of necessity by productively using the high drag on their wings. Hovering dragonflies, for instance, actually support a significant portion of their weight with their wings' drag as well as their lift (see Chapter 8).

As the wings of flying animals or aircraft get bigger, different properties change at different rates. Most insects use their wings at low Reynolds numbers, where wings are inherently less effective. At Reynolds numbers of about 7,000 (that of a large dragonfly or a hummingbird), wings become noticeably more effective; and changes in streamlining and airfoil shape start to play a bigger role.[27] Simple geometry also comes into play. If you take a bird-sized wing, maintain the shape, but double its wingspan, then obviously it will produce more lift. Having four times more area, it will, at the same flying speed, produce four times more lift. But weight increases with volume (or length cubed), so doubling the wingspan will make the wing eight times heavier. Doubling the wing span gives us four times more area but eight times more weight, so this wing's wing loading (or weight divided by wing area) has actually doubled. Because speed is related to wing loading, the wing will fly faster and actually produce more than four times more lift. So if the shape and the composition of a flyer remain the same as the flyer gets bigger, it will fly faster and its wings will produce disproportionately more lift.

For example, a square inch of dragonfly wing supports about 1/2,000 of a pound of the dragonfly's weight. In contrast, one square inch of a 747 jumbo jet's wing supports, on average, about 3/4 of a pound of the airplane's weight, more than 1,500 times greater than the dragonfly's wing. Wing loading usually increases dramatically with size. In standard metric units, a grasshopper has a wing loading of 3.5 newtons per meter squared (N/m^2); a bird such as a pigeon has 39 N/m^2; a small, four-seat, light airplane has 730 N/m^2; a medium-sized jet airliner has 5,300 N/m^2; and a jumbo jet has 6,600 N/m^2. In short, the jumbo jet carries the most weight with the least wing, giving it the most efficient wings.[28]

I fly radio-control (R/C) model airplanes, which are fun to consider in this context because they bridge the gap between birds and full-sized airplanes. The smallest common R/C models have wingspans of about fifteen inches, roughly the same as a robin's. Specialized indoor R/C and rubber-band-powered free-flight models get even smaller, with wingspans of ten inches or less. At the other end of the spectrum, a friend of mine has a scale model of a DC-3 airliner with a twelve-foot wingspan, and some of the largest scale models have wingspans of fifteen feet or more, half the wingspan of a Piper Cub.

Radio-control pilots who fly a wide size range of models soon learn from experience that wings change efficiency with size. For example, for models with wingspans of fifty inches or so, those with wing loadings of 45 N/m^2 (or 15 ounces per square foot to a U.S. modeler) are docile floaters and good for training. In contrast, a model with a wing loading of 66 N/m^2 will be fast and heavy and will require skill and experience to fly and land. If we move up to larger models (those, say, with seventy- or eighty-inch wingspans), one with a wing loading of 75 N/m^2 might be a reasonable trainer. If we look at smaller models, however, one with a twenty-inch wingspan and a wing loading of 45 N/m^2 would be a "lead sled" and quite challenging to fly. A large model with an eighty-inch wingspan and 75 N/m^2 wing loading can fly smoothly and slowly (for its size), but a small model with a twenty-inch span and a wing loading of

45 N/m^2 has to fly very fast to produce enough lift, making it a hot rod rather than a trainer. Newcomers to R/C modeling are often confused by these differences, mistakenly assuming that all trainers should have the same wing loading, whether they are dainty twenty-five-inch park flyers or ten-foot, quarter-scale behemoths.

Wing Structure

Until recently, certain aspects of conventional airplane spar-and-rib wing structures had not changed much since the era of the Wright brothers. The Wright Flyer had a wooden framework, with wooden spars running from wingtip to wingtip and airfoil-shaped ribs running from front to back. The whole structure was covered in cotton fabric (not the linen of later biplanes). During the 1920s and 1930s, wooden frameworks began to give way to metal—first the spars and then the ribs. Finally, thin aluminum-alloy sheets replaced the fabric covering. Although the aluminum skin panels weighed more than fabric did, the aluminum skin actually helped form part of the rigid structure of the wing; so the internal framework could be lighter and still produce a stronger wing.[29] Designers could also shape metal-skinned wings more precisely, allowing them to produce better airfoils. Lastly, metal-skinned wings could, in principle, be very smooth to reduce drag.[30] This was state-of-the-art airplane construction at the beginning of the Second World War, and it is still commonly used today. The major difference now is that for medium-sized and large subsonic airplanes (airliners and cargo transports) some of the skin and even parts of the framework may be lightweight composites rather than metal. Light single-engine and twin-engine airplanes still largely use the aluminum framework and skin construction, although some manufacturers are starting to use composites even in that sector.

The skin-over-framework structure lends itself nicely to carrying things inside the wing. Most ubiquitous are the cables, hydraulic lines, electrical wires, motors, and control rods and fittings for moving control surfaces. The next time you have a window seat on an airliner, watch the spoilers on the top of the wing when they pop up after landing. They expose the inside of the wing, and I am always amazed at the amount of plumbing and cabling visible in the relatively small space under the spoilers. Other common occupants of the space within the wing frame-work are fuel tanks and landing gear. During the Second World War, most fighter airplanes had their machine guns or cannons inside their wings so that the guns did not fire through the propeller disk. Jet fight-ers, however, with no propeller on the nose to worry about, usually have their guns in the fuselage or wing root. Most airplanes have marker lights and landing lights on the wings, so the wings must contain the wiring for the lights. Likewise, various sensors, such as pitot tubes for sensing speed (see Chapter 6) or tabs or slots for sensing approaching stalls, are usually on the wings, so tubes and wires for the sensors also run through the wings.

There is a major exception to the conventional framework-and-skins construction, and it is used by builders of supersonic military airplanes that need exceptionally strong wings. In at least one case, a manufac-turer has machined major wing sections out of solid blocks of steel or titanium, carving out cavities rather than riveting skins to a framework. Companies building the most advanced modern fighters today achieve much the same result with composites. Rather than carefully and tediously (and wastefully) machining away the unwanted parts of a block of metal, builders stack layers of the composite fabric to form the exact shape needed. Once it is heat-cured, the fabric integrates the skin and framework into a lightweight, unitary structure.[31]

The only other exception to these types of wing structures that I have found is the one used by the famous maverick designer, Burt Rutan, for his Vari-Eze line of homebuilt airplane kits. His method starts with a huge block of Styrofoam as long as a wing. The builder uses a heated wire to slice a precise wing shape out of the foam and then covers it with layers of fiberglass.[32] Other designers have adopted this technique for their kit-built designs. The advantage of fiberglass-over-foam is that it lets a builder produce a precisely shaped, smooth wing without having any metal- or woodworking skills. A disadvantage is that it is not a partic-ularly lightweight method, and the builder needs to be careful not to use too much resin; otherwise, the wing becomes downright heavy. Moreover, builders cannot put fuel tanks inside these wings; and adding lights, antennas, or wiring for controls or sensors is very difficult after the wing has been covered with fiberglass.

The wing structures of birds, bats, and insects are nearly as different from each other as any animal wing is from an airplane wing. We have already looked at insect wings with their pleated vein-and-membrane structure. An insect wing is essentially a two-dimensional structure that gets its thickness from pleating or corrugation, so it has virtually no internal space to store anything. True, the supporting veins are hollow and contain blood passages and air tubes, but the veins are only a fraction of a millimeter in diameter. They don't represent a significant volume compared to the wing size.

Insect wings are fundamentally different from the wings of all other flying animals because they have no muscles in the wings themselves. All changes in wing shape are produced by muscles acting on the wing joint (or parts of the exoskeleton attached to the joint).[33] The insect wing joint is a remarkably complex affair. Not only, in most species, does it allow the wings to fold back out of the way, but it also allows for slight differences in movement among different veins in the same wing. These movements allow the insect to do quite a bit of twisting and camber adjustment. Curiously, insect wings are made up of a composite material, the same stuff that makes up their exoskeleton, consisting of layers of microscopic fibers of chitin (a carbohydrate-like molecule) embedded in a protein matrix.

The rest of the flying animals are all vertebrates, and their wings are all modified front limbs. In fact, they are modified versions of our arms and hands; so they contain bones, muscles, nerves, and blood vessels, just as our arms do. Yet even with these basic similarities, the differences among the wings of the three major groups are dramatic.

Most of the surface area of a bird's wing is made up of feathers (fig. 2.5). Although feathers have a considerably more complex structure than hair does, feathers, like hair, are nonliving outgrowths of skin. They are mostly made of a waxy protein called keratin. If you took a photograph of an outstretched bird wing from above, you would see that well over half of the wing area is composed of feathers. If you poked a needle through most of the wing, you would not pierce

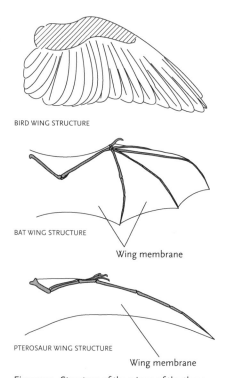

BIRD WING STRUCTURE

BAT WING STRUCTURE

Wing membrane

PTEROSAUR WING STRUCTURE

Wing membrane

Figure 2.5 Structure of the wings of the three vertebrate powered flyers. The crosshatched area of the bird wing contains the flesh-and-bone part; the rest consists only of feathers. For the bat and pterosaur, the wing membrane is outlined, and the bones are shaded.

any flesh, just feathers. The bird's arm and hand extend only about halfway to the wingtip, and typically they do not even come halfway to the trailing edge. The hand bones are reduced and fused into a single main strut supporting a single functional finger.[34]

Bird arms and hands have a unique, automatically extending structure: when the bird straightens its elbow, a special set of tendons also straightens its wrist and finger and extends and spreads the primary feathers (the large pinion feathers of the outer part of the wing). This mechanism cuts down on the number and size of muscles needed beyond the elbow, lightening the wing and making it easier to flap.[35]

Feathers are remarkably light, strong structures. The contour feathers that streamline the body and make up the wing surface have a central shaft with hundreds of small side branches called barbs. The barbs in turn have their own microscopic side branches called barbules, and the barbules have hooks that interlock with their neighbors almost like Velcro or zippers. These interlocking hooks hold each barb in its proper location to form a smooth surface. This is a handy failsafe structure because, if something pokes through a feather and disrupts the barbs, the bird can easily preen the feather with its beak to rearrange the barbs back to normal. The shapes of feathers are also well tailored to support flight loads. Each primary feather is shaped like a little cambered wing, and the shaft is always about one-third of the way back from the front edge of the feather, just as the main spar of an airplane's wing is usually about one-third of the way back from the wing's leading edge. A feather with this structure is well suited to support aerodynamic loads by itself and can also form a large, smooth surface when held alongside other feathers or partly overlapping them.[36]

One reason explaining why feathers are so light for their strength is that they are not a solid surface. Under a microscope, a feather looks like it has a surprising amount of air space: a gap of a similar size to the barb itself separates each barb from its neighbors. What looks like a solid feather to my naked eye may actually be about three-fourths feather and the rest air. Fortunately for the bird, the air also treats this clever arrangement as a solid surface. The gaps between barbs are much smaller than the thickness of the boundary layer, so no appreciable airflow can penetrate the barbs. In other words, feathers have it both ways. They have many tiny gaps to reduce their weight, but the gaps are so small that the air treats feathers as solid surfaces as it flows over a bird's wing.[37]

A bat's wing is also a modified front limb, but bats have evolved a structural arrangement that is almost the opposite of birds'. Whereas the flesh-and-blood arm structure of a bird's wing makes up relatively little of the wing's surface, the arm, hand, and particularly the fingers of a bat have been enormously expanded. The hugely elongated fingers support the outer half of a bat's wing, similar to the primary-feather region of a bird's wing. The arm and hand skeleton forms a framework, which the wing membrane stretches across (fig. 2.5).[38] Biologists long assumed that the membrane was just a somewhat stretchy layer of skin, but now we know that it contains precisely oriented elastic and reinforcing fibers as well as embedded muscles that can adjust the membrane's tension.[39] Although they look ridiculously spindly to us, the fingers of a bat's wing are quite mobile. They can substantially modify the wing's shape, particularly its camber and, to a lesser extent, its area.

How mobile are those wing-hands? Many bats "hawk" insects; that is, they pursue and catch flying insects while in flight themselves. In ideal circumstances, a bat's pursuit of an insect ends with the bat's jaws closing on the prey. Circumstances are not always ideal, however, and last-instant evasive action can take the insect quarry just out of reach of the bat's mouth. When this happens, most bats can use their wings to catch the insect, either by cupping the wing to catch and hold the insect against the membrane while flexing the wing to bring it to the bat's mouth or by striking the insect with a wing to deflect the insect to the bat's mouth. Doing this in flight is a very impressive feat.[40]

The extinct flying reptiles called pterosaurs (or sometimes ptero-dactyls) had structures more like bats' wings than birds': elongated arm and finger bones supporting a membrane. Unfortunately, with only fossils to go on, we don't know as much about pterosaur wing structure as we do about bats', but some features are obvious even from fossils. As in bat wings, most of the wing surface is a membrane supported by a modified arm and hand skeleton. Unlike bats, however, pterosaurs supported their wing membrane with arm bones and a single, enormously elongated finger. *Pterodactyl* in fact means "wing-finger." We know that this giant finger supported a flexible membrane that ran from the fingertip to the side of the animal's body because several fossils show impressions of the membrane.[41] What the fossils don't show is what the membrane was made of. Was it stretchy like a bat's wing, or stiff like a modern sailboat's sail? Did it billow freely like cloth, or did it have stiffeners to give it a fixed camber? Although the traditional view is that the membrane was bat-like, at least one paleontologist has argued that grooves in the impressions of some wing membranes show that the membrane was actually covered with stiffening fibers, making it almost as rigid as a feathered bird's wing.[42] Scientists need more research (and proba-bly more fossils) to resolve such questions, but we may never know the details of how pterosaur wings worked.

Other Uses for Wings

Clearly, the main reason to have a wing is to fly (or, if you are an ani-mal such as an ostrich, because your ancestors flew). But wings, both animal and airplane, also have a number of other uses; and among animals, some of those secondary uses are stunning.

Airplane wings are always primarily lifting surfaces, though I have already discussed secondary functions such as storing fuel and supporting lights and instruments. Many airplanes also have landing gear on the wings, either attached externally and fixed or retractable so they can be contained inside the wing. Most multi-engine airplanes have their engines mounted on or in the wings (see Chapter 3). Yet even though these secondary functions can be of great importance to particular types of airplanes, a number of airplane wings have few or none of these extra functions—for instance, the wings of early models of the Piper Cub (from the 1940s) and those of most ultralight or light sport airplanes of today. Moreover, all of these secondary functions are directly related to flight: fuel for power, lights and instruments for safety, and landing gear for landing.

Animals are different. Many animals use their wings for purposes totally unrelated to flight. Insects, for example, commonly use their wings for protection from predators. Wings can provide camouflage: consider moth wings that blend in with tree bark or bright green katydid wings that look like leaves. Or they can repel predators more actively. Many moths and butterflies have large, bright eyespots on their hindwings that look like the eyes of an owl or another vertebrate predator. Most of the time the insect keeps the eyespots covered by the forewings; but if a bird gets too close, the insect quickly flashes the eyespots. The sudden appearance of what looks like a potentially dangerous predator is often enough to startle the bird and cause it to fly off and look elsewhere for a meal (see Chapter 9). One of the most amazing displays is used by a group of small flies. Tephretid flies have what looks at a glance like a splotchy dark pattern on otherwise clear wings. These flies must have been a common prey of jumping spiders over the millennia because they have evolved an anti-spider defensive display. When a fly sees a jumping spider, it quickly faces the spider and holds out both wings rotated vertically so that the lower wing surface faces the spider. Then it waves the wings up and down. The pattern on the wings, combined with the fly's body, form an outline that looks to the spider like another jumping spider making a threat display. Because jumping spiders have fairly good vision (as spiders go) and are antisocial—to the point of eating each other if they get too close—a jumping spider faced with a tephretid fly display usually backs off and hunts elsewhere.[43]

Many birds—the killdeer is probably the best-known example—use the broken-wing display. If a large animal that might be a predator gets too close to a killdeer sitting on its nest, the adult bird gets up and runs away from the nest, trailing one wing at an awkward angle as if injured. It then stops and waits for the predator to approach before running off for a short distance. The killdeer repeats this display, leading the predator away from the nest while never allowing it to get close enough to be dangerous. Once the predator is well out of sight of the nest, the killdeer miraculously recovers from its "broken" wing and flies back to its nest.

Bearing out the old saying that the best defense is a good offense, geese use their wings as weapons. A lump of thickened bone forms on the leading edge of a goose's wing, which the goose can use as a club. City folk are often amused to hear that rural people use geese as guard animals, but they would not be so amused if they were to become the center of attention of a flock of angry geese. In addition to providing a noisy alarm, geese use their wings to beat off predators. Several geese working together have been known to drive off foxes and dogs.

Many flying animals use their wings in various ways to promote mating and reproduction. Some insects, especially some butterflies, have brightly colored wings that they use to attract mates. Picture-wing fruit flies use semaphore-like wing waving as a dominance display between males or as a courtship display to attract females.[44] Animals as varied as fruit flies, crickets, and South American birds called manakins make wing noises (when not flying) to attract mates or as part of a courtship ritual.[45] A few species of crickets have one of the more bizarre wing adaptations: the males' wings are thickened and filled with nutritious protein. When the crickets mate, the female eats the male's wings during copulation. Although closely related cricket species can fly, and presumably their ancestors flew as well, males of these crickets cannot fly (the wings are too short and thick), and the females are completely wingless.[46]

Bees have evolved several secondary wing functions. Large bees such as bumblebees often collect pollen from flowers of plants like tomatoes and eggplants that require "buzz pollination." The bee pushes its body up against the flower's anthers (pollen-containing structures) and rapidly vibrates its wings. The vibrations cause the anthers to eject a small amount of pollen, which the bee collects to feed its larvae. Honeybees, with their large, highly organized colonies, have several mechanisms to keep their hives at a particular temperature. If the hive gets too warm, many of the workers in the hive will start flapping their wings while holding onto the comb. With dozens or hundreds of bees fanning their wings this way, they can push lots of air through the hive and cool it down effectively.

Birds are bipedal and get around on the ground without using their wings. But bats are still largely quadrupedal and need both their wings and hind legs for terrestrial locomotion. Bat wings make rather awkward front legs, so most bats are not very agile on the ground. They do tend to be excellent climbers, however. They have retained a more or less normal-sized clawed thumb separate from the wing membrane, which they depend on heavily when climbing. At least one kind of bat, however, the vampire bat, can walk fairly well on the ground.[47] Unlike other bats, it approaches its prey (large sleeping mammals such as cows or deer) by creeping up on the ground. The vampire bat bites a small cut through its victim's skin and laps up the blood. Vampire bats typically bite the victim's feet or legs from the ground, but they can also climb onto the victim's body if necessary.

The Two Big Differences

Although they depend on the same basic aerodynamic processes and are generally shaped by the five features discussed previously in the chapter, animal wings and airplane wings have two fundamental differences. One is the underlying processes that shape or modify the wings: evolution by natural selection for animals, engineering for airplanes. The other is how they are used: animals flap their wings to produce thrust as well as lift; airplane wings are fixed, and a separate engine produces thrust.

Why do animal wings have so many more secondary functions than airplane wings, including some functions that interfere with flying? Consider the goals of a flying animal versus an airplane. The most important goal of any animal is to reproduce, to have its genes represented in the next generation. Natural selection acts on the whole animal and all of its biology. If a species encounters new conditions in which some individuals reproduce better if they become better runners at the cost of becoming worse flyers, then most likely the whole species will eventually become better runners and worse flyers. Animals only retain the ability to fly as long as it gives them an advantage. If conditions change so that a flying species is more successful without flying, chances are that it will eventually lose the ability to fly.

In contrast, the goal of an airplane is to fly, whether for training, recreation, transport, or combat. Engineers design airplanes to fit a particular mission, but that mission always involves flight. A flightless airplane is an oxymoron; I can conceive of no practical reason to build an airplane that does not fly. Thus, all secondary functions for airplane wings are related more or less directly to flight. Instruments and lights make flight safer, fuel tanks make flights longer, wheels make landings easier. Airplanes (and their designers) don't need to worry about traveling as easily on the ground as in the air, or fending off predators, or attracting mates. Designers just need to worry about how effective an airplane will be at fulfilling its aerial mission.

The other basic difference between animals and airplanes is operational rather than goal-related. Airplane wings are largely static compared to animal wings. Airplane wings do not flap, they do not radically change shape, and (except for a few specialized types) they do not fold up for stowage. This largely static arrangement allows designers to use simpler, lighter attachment structures and also allows them to tailor wing shape to optimize aerodynamic performance for a particular mission: training pilots, carrying cargo, flying aerobatics at air shows, and so on.

Animals' wings are inherently dynamic and flexible, both literally and figuratively. Animals need to flap their wings, so their wings must be highly mobile and adjustable. The flexibility necessary for flapping places design constraints on a wing, but it also opens opportunities. All that adjustability gives flying animals tremendous ability to change directions quickly, so they are much more agile than any airplane. This flexibility also translates into foldability for most flying animals, who can get their wings out of the way when not in use.

Getting those wings folded out of the way is nothing to sneeze at. Imagine a honeybee hive if all the bees had to clamber around with extended wings or a seabird rookery on a cliff or a bat roost in a cave if all the thousands of birds or bats had to walk or hang around with fully outstretched wings. Now consider airplanes. Imagine how much smaller airport terminals or maintenance hangers could be if airplanes could fold their wings and park shoulder to shoulder. And consider the thousands of acres of land devoted to aircraft bone yards. Thousands of perfectly good surplus or obsolescent airplanes are parked at desert locations such as Davis-Monthan Air Force Base or Mojave Airport. These airplanes are not needed right now but are too valuable to scrap, so they are parked and mothballed, just in case. Imagine how much less space they would take up and how many more places they could be stored in if they did not have wings sticking out thirty or fifty feet on each side. As aerodynamically efficient as those airplane wings are, they can become a significant headache when the airplane is on the ground.

~ 3 ~

Power

The Primary Push

Contrary to popular belief, if an airplane's engine stops, the airplane does not immediately fall out of the sky and crash. In 1983, an Air Canada Boeing 767 lost all engine power when flying at 28,000 feet above sea level. (The airliner had run out of fuel due to a malfunctioning automated fuel-control system and confusion among the pilots and ground crew about whether to use English or metric units for manually calculating fuel needs.)[1] Entirely without engines, the pilots were able to glide the airliner more than ten miles to a more or less safe landing at an abandoned Canadian Air Force base, much to the amazement of people picnicking and racing cars on the abandoned runway. Oddly enough, the only passenger injuries (a few sprained and broken ankles) occurred as they were exiting the airplane. The airplane was repaired and back in service in a few months.

In a glide, a flyer can still fly; but without a power source, a flying animal or an airplane is unable to fly at a level altitude or to climb. It must descend. A flyer's power system uses the energy in fuel to produce thrust. Because wings cannot produce lift unless they are moving through the air (see Chapter 2), the power source keeps the flyer moving so that its wings work regardless of whether it is going up or down or staying level.

Power Sources: The Great Divide

The most fundamental difference between flying animals and flying machines is their difference in power sources. All flying animals use muscles; all practical airplanes use internal combustion engines. These two power sources are based on completely different operating principles. This difference has a huge effect on their structural arrangements, control mechanisms, and size limits.

Animals, from the tiniest mosquito to the largest pterosaur, power their flight with muscles. We, too, are powered by muscles, which we tend to take for granted as long as they work properly. Muscles are quite unlike the engines or motors that power most of our machinery. An engineer might describe a muscle as a unidirectional, linear actuator of limited stroke. Muscles are linear actuators because they work by getting shorter. Imagine taking an individual muscle, tying one end to a hook on the wall, and hanging a weight from the other end. When you activate the muscle, it shortens, pulling the weight toward the hook. This is a linear motion and unidirectional because the only active movement the muscle can make is to get shorter. The weight will stretch the muscle back out when the muscle relaxes, but the muscle cannot actively lengthen. Without the weight, it would just stay short. And most muscles don't shorten very much. From a normal relaxed length, typical vertebrate muscles might shorten by 10 or 15 percent, whereas some insect flight muscles shorten only by 3 or 4 percent—a limited stroke indeed.[2]

Suppose I want to hold my arm out in front of me, then swing it out to the side, and then move it back in front of myself. If muscle works in only one direction, how can I move my arm back and forth? Muscles generally work in pairs, and these muscle pairs need the joints (or articulations) of a skeleton. Take the classic example of my elbow. I have a muscle on the front of my upper arm (the biceps) that bends, or flexes, my elbow. I have another muscle on the back of my upper arm (the triceps) that straightens, or extends, my elbow. These two muscles are called antagonistic muscles because each reverses the action of the other. What many people don't realize is that the joint is as important as the muscles. If an injury has locked my elbow and prevented me from bending it, then either muscle can pull for all it's worth but won't produce any movement. In short, the most basic movements require a pair of antagonistic muscles and a joint. Many joints can move in more than one direction—our shoulder joints, for example—so they have more than one pair of antagonistic muscles operating on them.

All flying animals—birds, bats, pterosaurs, and insects—have rigid, jointed skeletons. With muscles for motors, animals with rigid skeletons are limited to see-saw or to-and-fro types of movements at joints. In other words, one muscle contracts to move the limb as far as it can in one direction, and then the antagonistic muscle contracts to move the limb back in the opposite direction.[3] When I walk, one set of muscles swings my leg back to push me forward during the power phase of a step; and a different set swings my leg forward during the recovery phase to prepare for the next step. When a flying animal such as a honeybee or a sparrow flaps its wings, one set of muscles moves the wing down during the downstroke, and a separate antagonistic set moves the wing back up during the upstroke.

What do these muscle movements have to do with power? To a physicist or an engineer, power is force times speed, and muscles exert forces that cause appendages or whole bodies to move at some speed. If I walk up the stairs, my leg muscles push down on the stairs with some force, and I go up the stairs at some speed. Running up the stairs is faster than walking up, so running up stairs requires more power than walking up. When a bird flaps its wings, the bird's wing muscles generate a force to move the wings through the air, which in turn pushes the bird through the air at some speed. In principle, multiplying that force times the speed ought to give the bird's mechanical power output, which sounds simple. But knowing which component of the speed to use and measuring the force the bird exerts are real challenges. Scientists have attempted to meet these challenges by using metabolic power input or muscle mechanics (which require simplifying assumptions about efficiency) or aerodynamic calculations (which require simplifying assumptions due to complex flapping movements).[4] Using combinations of these approaches, they have measured power outputs of flying animals as diverse as bumble-bees, hummingbirds, parakeets, and falcons.[5]

Engines, rather than muscles, provide the power for airplanes. The engines that power airplanes come in a huge variety of forms, but all practical airplane power plants are internal combustion engines: they burn a petroleum-based fuel inside the engine and use the heat released to produce power. The three basic engine types are piston engines (also called reciprocating engines due to the to-and-fro movement of the pistons), pure jets (including turbojet and turbofan engines), and turboprop (or propjet) engines.[6] Piston engines work just like everyday automobile engines, with the pistons turning a crankshaft that turns a propeller instead of wheels. The piston airplane engine reached its zenith around World War II, with some aero engines dwarfing even the most powerful truck engines in size and power. Ironically, due to the conservatism imposed by the emphasis on safety and reliability in civilian aviation, technological progress in piston airplane engines tapered off after the war, lagging far behind developments in piston engines for automobiles. Today's typical small airplane engine has a carburetor with a throttle and a separate, manual mixture control as well as an electromechanical magneto to drive the spark plugs. For at least a decade, automobiles have had computer-controlled fuel injection with automatic mixture regulation and electronic (solid state) spark plug ignition with computer-adjusted timing. (Magnetos were commonly used on early automobiles such as the Ford Model-T because they can produce the spark to start an engine without a battery; aero engine manufacturers have long regarded that ability as a safety feature, so they have retained these primitive devices for many decades after the auto makers abandoned them.) More advanced piston engines are becoming available for airplanes—for a price—but basic single-engine airplanes are still largely powered by 1930s engine technology.

World War II also saw the introduction of jet engines.[7] Pure jet engines work by burning fuel to heat air in the combustion chamber, which increases greatly in volume and pressure and rushes out the tailpipe. The gas that is flowing backward produces Isaac Newton's "equal and opposite reaction," pushing the engine (and its attached airplane) forward.

Although engineers have come up with several different kinds of jet engines, all the practical ones used on airplanes are turbojets or turbofans. A turbojet uses the hot exhaust gas to spin a turbine (a very small, fast, high-tech windmill), which in turn spins a set of compressor blades at the front of the engine. The compressor sucks air into the front and compresses it, raising the temperature and pressure of the air as it arrives in the combustion section. Burning fuel in the combustion section further increases the pressure of the air before it passes through the turbine and out the exhaust. A turbofan is like a turbojet with one set of greatly expanded compressor blades at the very front of the engine. This first set of compressor blades blows back some of the air into the next set of compressor blades and on into the engine, but it also blows some of the air around the outside of the engine core. This cool bypass air produces some thrust, increases the engine's fuel efficiency at some speeds, and greatly reduces its noise. All modern jet engines are turbofans.[8]

A turboprop engine is a sort of hybrid propeller-jet engine. It is essentially a jet engine that uses a more powerful turbine to spin a propeller as well as a compressor. Turboprops have operating characteristics somewhere in between piston engines and pure jets, and they have largely replaced piston engines in commuter airliners and other small, high-performance aircraft. Figure 3.1 illustrates the three types of gas turbine engines.

Regardless of whether they use pistons or turbines, all aircraft engines are based on rotating motion: one or more shafts rotating continuously in one direction. Even a piston engine converts the reciprocating motion of its pistons into a continuously rotating propeller shaft. Machinery as a whole is based strongly on rotating parts—shafts, axles, pulleys, gears, wheels. More than a phenomenon of the Industrial Revolution, Archimedes-screw pumps and cartwheels have been around for thousands of years, windmills for more than a millennium. Whether the force comes from steam, electricity, internal combustion, or even a domestic animal's muscles, we have a

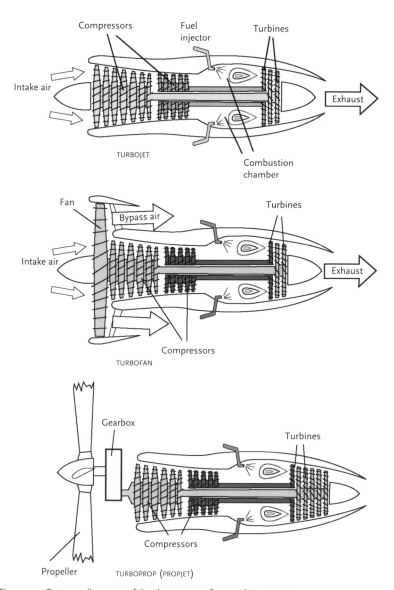

Figure 3.1 Cutaway diagrams of the three types of gas turbine engines.

long tradition of converting that force into rotating motion to power our devices and mechanisms.

Our fondness for rotating machinery is not just a quirk or a random bias. A continuously rotating device is, all else being equal, more efficient at power transmission. If a device has a back-and-forth action with a power and recovery stroke, such as oars on a rowboat, the device is applying force only about half the time. Consider my bicycle tire pump: the only time I usefully force air into the tire is when I push the handle down. When I pull the handle back up, I am not adding any air to the tire. Even if I could apply power to a back-and-forth mechanism on both strokes, it still would be less efficient than continuous rotation because of the need to decelerate and stop and then accelerate in the opposite direction at the end of each half stroke. When two children play on a teeter-totter, the child on the down-going end must absorb the energy and decelerate the seat with her legs before she can push off and raise her end. Her brother must do the same when his end goes down. If they were riding on a merry-go-round instead, their legs would always be pushing in the same direction, and they would not waste any effort in decelerating and reversing the motion. A rotating mechanism will normally apply power more effectively than will a back-and-forth or see-saw type of mechanism.

Do muscles and internal combustion engines have anything in common? Both muscles and engines get stronger as they get bigger. A Piper Cub's seventy-five-horsepower engine weighs 180 pounds and can produce just enough power to carry aloft two people and a light airplane weighing eight hundred pounds. A De Havilland Beaver's 670-pound engine produces 450 horsepower and can easily keep five people and a large load of camping gear airborne in a 3,000-pound airplane.[9] These airplanes are very similar in layout—strut-braced, high-wing, single-engine cabin monoplanes—yet the Beaver's bigger engine puts out a lot more horsepower and can fly a much larger airplane. By the same token, bigger animals have bigger muscles and can also carry heavier loads. A hummingbird might be able to carry a 1/10-ounce twig back to its half-built nest, whereas an eagle can easily carry a three-pound duck or rabbit back to a nest of hungry chicks.

Saying that muscles and engines both get more powerful as they get bigger masks a rather important difference. As engines get bigger, their specific power, or power output per pound of engine, changes some but not very much over a huge range of engine sizes. Muscles are different. Their force production—and, hence, power output—is related to their cross-sectional area. Because weight is strongly tied to volume, muscle power and weight follow a surface-to-volume or square-cube relationship. If an animal gets twice as long, all else being equal, its muscle cross-section will increase by two squared, or four times, but its weight will increase by two cubed, or eight times. So muscles do get stronger as they get bigger, but their power increase does not quite keep up with their weight increase. The converse is that, as animals get small, their muscles become more powerful for their weight, which is why ants can carry a load several times their own body weight. This disproportionately high power in small critters has important consequences for flight, as we'll see shortly.

When I first looked into engine power, I expected that, at least within major engine types, airplane engines would produce more power per pound of engine as engines got bigger, by a sort of economies-of-scale relationship. In fact, specific power does not seem to have a strong connection to size. I collected engine performance data for engines now in production from the online edition of the authoritative *Jane's All the World's Aircraft* to compare engines in a range of sizes.[10] For example, my selection of air-cooled, four-stroke piston engines included twenty-eight models from ten companies in nine countries. Over a range of engine weight from 120 to 1,300 pounds (more than a ten-fold increase), specific power measured as horsepower per pound ranged from 0.4 to 0.77 (less than a two-fold range). Moreover, the differences in specific power did not seem to be at all related to weight: the engine with the lowest specific power (0.4) weighed about the same, 160 pounds, as the engine with the second-highest specific power (0.75). The situation was much the same among a random selection of nineteen jet engines. With more than a four-fold range in engine weights (459 to 16,664 pounds), the thrust per pound of engine varied from 3.3 to 6.5 and, again, was not noticeably associated with size. In a nutshell, the power per pound of engine does not show any consistent change with engine size. To get a big jump in power per weight, you have to change to a different engine type: turboprops do better than piston engines, and jets do best of all.

Round and Round It Goes

One factor probably explains most of the fundamental difference in design between nature's flyers and flying machines: our machines are full of wheels and axles, but no animal has ever evolved a true axle or wheel with complete rotation. (No, contrary to folklore, owls cannot turn their heads all the way around.) Explaining why animals have not evolved something is always a dicier proposition than figuring out how some existing adaptation arose; maybe the right combination of selection pressures and appropriate animals has just not yet come together. In this case, however, two facts are telling. First, with unidirectional muscles attached at both ends to a skeleton, animals are unable to power a continuously rotating axle. Moreover, we are used to a world in which humans put money and effort into making smooth, relatively level roads. As my colleagues, Michael LaBarbera and Steven Vogel, are fond of pointing out, wheels may not be all that much use to an animal.[11] How much use would a car or truck be in a world with no roads? Or, to put the situation into terms more relevant to most animals, would you rather trek through rocky hills and dense forests on rollerblades or legs? In short, the selection pressure to evolve wheels has probably been nonexistent for animals in natural environments.

If you can't use axles and you use muscles for power, flapping your wings to produce thrust is a very effective solution. In contrast, if you can use axles, then you can use one small, specialized, spinning wing—a propeller—for thrust and keep your main wings fixed and just use them for lift. This difference is the basis for why all flying animals flap their wings yet no practical airplanes use flapping wings. Flapping-wing flying machines (called ornithopters) are not technologically impossible by any means. Many people have built radio-control ornithopters in the past decade (some are even available as kits), and at least one person-carrying ornithopter has made a brief flight (see Chapter 10). But even the developer of a full-sized ornithopter himself admits that it will have little or no practical impact on aviation.[12]

Muscles and engines also differ dramatically in the types of fuel they use and how it is consumed, although the efficiencies are not terribly different. Engines burn hydrocarbon fuel at very high temperatures and tend to produce considerable amounts of waste heat, but engine efficiency is in principle easy to determine: compare the energy in the fuel to the amount of work the engine does, and you get a direct measure of efficiency. Specifically, efficiency is the ratio of mechanical work done to energy consumed. Piston engines are about 30-percent efficient, jet engines somewhat more so. Animals are, as usual, much messier. They use carbohydrates, fats, and proteins from their food as fuel and oxidize these molecules at room temperature using sophisticated catalysts. In principle, under optimal conditions animals can capture 30 to 35 percent of the energy in their fuel (food). Assessing the mechanical efficiency of a bird's flight motor is tricky, however, because not all the energy from fuel goes into flapping the bird's wings. Some goes into breathing, pumping blood, sending nerve signals, and even digesting food to obtain more fuel. Even if we could isolate just the flight muscles, measuring efficiency is challenging because biologists have problems measuring transmission losses (such as bending joints, stretching tendons and skin) as well as exactly how much aerodynamic work the wings do while flapping. Best estimates so far are that flight muscles in birds and insects are in the neighborhood of 14- to 20-percent efficient; that is, only 14 to 20 percent of the animal's metabolic energy use appears as mechanical work.[13]

Flapping Wings: Masters of All Trades?

Although the power system of flying animals may be inherently less efficient than having a separate propeller for thrust, animal wings are also vastly more versatile than airplane wings. Animals can move the wings, change their shape, twist and bend them, and, in many cases, even change their size. Small and medium-sized flying animals can hover, and all but the largest can take off and land vertically. Flying through a forest without hitting any branches or trunks is all in a day's work for a sparrow or a bat or a butterfly. Moreover, the vast majority of flying animals can fold their wings out of the way and get around completely on foot as well. In spite of many decades of effort, engineers have yet to develop an airplane that also works as a decent road-going vehicle, let alone going off-road.

The basic flapping motion is surprisingly similar for all flying animals, from house fly to sandhill cranes. The wings don't simply move up and down like a door turned on its side. On the downstroke, the wing moves down and forward with the leading edge tilted slightly down. Tilting the leading edge down tilts the lift forward, which produces thrust. The downstroke is generally the main power phase of the stroke, when the wing produces lots of lift and thrust. At the bottom of the downstroke, the flyer sharply tilts its wing nose up and sweeps it up and back. The upstroke varies depending on the kind of animal. The wings of large birds may produce some lift but no thrust during the upstroke, whereas many insects and some small bats appear to get extra thrust but no lift on the upstroke.[14] For most moderate-sized flyers (large insects, small and medium-sized birds), the upstroke is a recovery phase that contributes little or no lift or thrust. Regardless of size, the majority of the useful forces are always produced during the downstroke. This difference between the upstroke and downstroke is often reflected in the muscles. For example, in chickens and turkeys, the so-called breast meat is actually flight muscle. The downstroke muscle is the deltoid—the larger, thicker, outer part. The upstroke muscle is the supracoracoideus, a much smaller, deeper part, which accounts for much less than half the mass of the deltoid.[15]

In addition to the five features of highly effective wings discussed in Chapter 2 (surface area, aspect ratio, orientation, structural strength, and airfoil shape), wings meant for flapping need a few other properties. First, a wing needs to join the body at a flexible joint or articulation rather than at a fixed root. This joint requires rather contradictory properties. It must be mobile enough to allow the wing to flap but strong enough to carry the animal's weight. When the animal is in flight, its entire body weight is essentially hanging from those wing joints. A rigid wing root (as on an airplane) is relatively easy to strengthen, but a flexible joint is more challenging to make strong while still allowing for mobility. Smaller, lighter flyers such as hummingbirds can get away with very mobile, flexible shoulders, whereas large birds have shoulders that allow for basic flapping movements and little else.

Besides the up-and-down wing motion of flapping, another key motion is rotation of the wing around its long axis—that is, tilting the wing leading-edge-up or leading-edge-down. Biologists call tilting the leading edge down *pronation* (from *prone,* or lying face down) and turning the wing leading edge up *supination* (from *supine,* or lying face up). Pronation and supination are critical movements for thrust production in animals. If birds could not pronate their wings on the downstroke and supinate them on the upstroke, they would produce no thrust, and flapping would be futile. If I need to pronate and supinate my wing while also sweeping it up and down, I can use either a ball-and-socket joint, like the shoulders of birds and bats, or a multipart hinge, as insects do. The insect wing joint consists of at least two separate hinges that allow the front and back of the wing to go up and down in sequence rather than simultaneously.[16]

Birds, bats, and insects can all change the shape of their wings in ways that airplane designers can only envy. Birds, for example, can collapse their wing feathers together like we can raise Venetian blinds, thus cutting their wing area in half or more, a potent advantage for quick descents. Insects, having no muscle in the wing itself, cannot significantly change a wing's area in flight. Their sophisticated multi-hinge joint, however, lets them do lots of twisting and bending, changing from a wing as flat as a billiard table to one as convex as the roof of a Quonset hut. Moreover, all birds and bats and most insects can fold their wings compactly out of the way when not flying, so they can walk (or climb or burrow or swim) as easily as they can fly.

In spite of this mobility and flexibility, animal wings are, after all, still wings. They need to be as light as possible, both for ease of flapping and efficient flight, while still being strong enough to support the animal's weight in flight. The clever, pleated, vein-and-membrane structure of insect wings (discussed in Chapter 2) yields very stiff structures for their size. Feathers, too, are surprisingly stiff and strong for their weight, particularly in the directions useful for supporting aerodynamic loads. So for their size, animal wings are very strong.

Divide and Conquer: Engines for Thrust, Wings for Lift

Of the three basic types of practical airplane engines, piston engines have been around the longest. The engine that powered the Wright Flyer in 1903 is based on the same principles as the engines that powered famous World War II airplanes such as the P-51 Mustang fighter and the B-17 Flying Fortress bomber as well as the Cessna 172 trainer down at your local airport today. Piston engines come in many varieties, but the typical piston airplane engine is a variation on the typical automobile engine: a four-stroke gasoline engine. Over the years, engine designers have built airplane engines with as few as two and as many as thirty-six cylinders in in-line, vee, flat (or opposed), and radial cylinder arrangements. Some airplane engines, including the famous Rolls Royce Merlin (used in Mustang and Spitfire fighters and Lancaster bombers) are water-cooled, just like a typical car engine. Other airplane piston engines are air-cooled, which works because airplanes fly much faster than cars can drive and therefore benefit from a greater amount of cooling airflow.[17] The biggest piston engines were developed during and shortly after World War II. The large, powerful Pratt and Whitney R-2800 eighteen-cylinder radial engine was used in fighters, bombers, and cargo airplanes. It had two banks of nine cylinders each so that it looked like two nine-cylinder radial engines merged front to back. The Pratt and Whitney R-4360 was a twenty-eight-cylinder engine, the largest piston engine built in the United States in significant numbers, and it powered a variety of large bombers and cargo airplanes. It had four radial banks of seven cylinders each and produced about 4,000 horsepower.[18] Lycoming built prototypes for a monstrous thirty-six-cylinder, 5,000-horsepower engine near the end of the war, but by that time engineers had begun to realize that jet engines were the wave of the future for high power needs. The huge Lycoming engine never went into production.[19]

A piston airplane engine does its job by spinning a propeller, which is actually a small wing that produces its lift parallel to the propeller shaft. In other words, the propeller's lift is aimed forward and represents the thrust of the engine. Propellers have some advantages over jet engines. For example, they respond more quickly to throttle changes, and the slipstream they blow over the tail surfaces can improve low-speed handling. Propellers have a fundamental speed limit, however. As the airflow over the blades approaches the speed of sound, the blades produce less lift (hence, less thrust) and require a huge increase in power—more power than can be produced by any practical piston engine. Because a propeller is spinning as well as moving forward with the plane, air is always moving past the propeller much faster than it is moving past the whole airplane. This is especially true at the blade tips. The propeller blades thus approach the speed of sound well before the rest of the airplane, to the point that a propeller-driven airplane simply cannot go much faster than about five hundred miles per hour, about three-fourths of the speed of sound at sea level. The only way to go faster is to use some other kind of engine.

In 1931, British engineer Frank Whittle patented a design for a turbojet engine for use in airplanes, but he could not convince anyone in Great Britain to develop the engine or put it into an airplane until after the beginning of World War II.[20] The Germans were quicker to see the advantage of the jet engine. Working independently from Whittle and starting later, they developed jet engines to the point of being able to field at least two jet airplanes—the Me-262 fighter and the Ar-234 reconnaissance bomber—long before the Allies had any practical jet aircraft. By the end of World War II, engineers and airplane designers realized that the most significant increases in speed and power were going to come from jet engines. Jet engines do not have the fundamental speed limit of a propeller and can be designed to operate well above the speed of sound.

The early jet engines were built using technology near the frontiers of engineering knowledge. Their turbines and compressors spun at much higher speeds and operated at higher temperatures and pressures than piston engines did. The first generation of jet engines was prone to flameout (complete loss of engine power) from several causes, including the act of simply advancing the throttle too fast. Because of their enormous increase in spinning speeds (many jet engines idle at higher revolutions per minute than the top speed of most piston engines), these first jet engines required much more frequent maintenance than piston engines did and were much less reliable. The trade-off was worth the early headaches, however. Even the first crop of jet engines produced more thrust than did a piston engine of similar weight; and with no propeller, they had no inherent speed limit. Today, even the humblest jet airliner cruises 30 to 40 percent faster than the top speed of the fastest piston-engine fighters of World War II. Moreover, jet engines become more efficient at high altitudes (more than 20,000 feet or so), where piston engines either perform very poorly or require complex superchargers.

Jet engine design has come a long way, and today's jet engines (now mostly turbofans) are actually much more reliable and require much less maintenance than piston engines do, largely because they have many fewer moving parts. They are also lighter, more powerful, and, at high altitudes, much more fuel-efficient. With all these advantages, why are some airplanes still using piston engines? Initial purchase price is probably a major factor. The tight manufacturing tolerances required by high temperatures and pressures produce high reliability but also give jet engines a relatively high price tag. Jet engines are also thirsty at low altitudes and airspeeds, where they lose their fuel efficiency. Consider a basic four-seat airplane with an unpressurized cabin and no supplemental oxygen system, such as a Cessna 172 or a Piper Cherokee. Without supplemental oxygen, the pilot cannot legally fly the plane above 10,000 feet, and I suspect that most such airplanes rarely fly even half that high. At the low altitudes and speeds of these small airplanes, jet engines may consume more fuel than piston engines do.

The turboprop engine might seem to combine the worst of both worlds: the speed-limiting propeller and the expensive jet. In fact, however, the turboprop engine fits nicely between piston and pure jet engines in several ways. Even with the speed limit, propellers have a number of advantages. They are more efficient than jets at low speeds because they accelerate a lot of air by a small amount; in contrast, jets accelerate a little air by a large amount. Pure jet engines also have a rather sluggish response to throttle changes, but turboprops can give more rapid speed-control response. All turboprops have propellers with adjustable pitch: the prop blades twist on the propeller hub so that the blades take a bigger or smaller bite of air, allowing for quick speed adjustments. Moreover, thanks to the efficiency of propellers at low speeds, turboprops are actually more fuel-efficient at moderate speeds and altitudes than are either pure jets or piston engines.[21]

Turboprop engines uniformly power the smallest modern airliners, the so-called commuter planes. In airline service, where airplanes are in almost constant use, long intervals between overhauls and lower maintenance costs more than outweigh the higher initial purchase cost of turboprops versus piston engines. Even without this economic benefit, airlines might still use turboprops because of their higher reliability, which is a safety advantage. Finally, turboprops produce about twice as much power for their weight as piston engines do. This difference further improves the airplane's performance (higher speed and greater payload) and thus its profitability.

Over the years, designers have used (or proposed) other types of engines in airplanes. In the earliest days of aviation, Samuel Langley, among others, built small flying models powered by steam engines; and some inventors, such as Hiram Maxim, tried unsuccessfully to use steam engines in full-size airplanes.[22] In the 1930s and 1940s, the Germans built some diesel-powered airplanes, but diesels tend to be heavy for their power output and have not caught on in aviation.[23] For many years, Rotax has been selling two-stroke gasoline engines for the ultralight and homebuilt market. Two-stroke engines tend to have higher power for their weight than four-stroke engines do, which is why they are used in chainsaws and many smaller motorcycles and outboard boat motors. Two-strokes also have many fewer moving parts so tend to be cheaper. Yet they are also much noisier and more fuel-thirsty, have dirtier exhausts, and are arguably less reliable. As a result, they have remained a niche market in aviation.

At the outlandish extremes, quite a few different airplanes have flown on rocket power, including the German Me-163 interceptor (more dangerous to its pilots than to the Allied planes they were meant to attack); the Bell X-1, in which Charles Yeager became the first person to fly faster than the speed of sound; and the North American Aviation X-15, the fastest winged airplane ever to fly: 4,500 miles per hour, faster than Mach 6.[24] Some air-racing enthusiasts are even promoting a new class of small, rocket-powered racing airplanes. Probably the most exotic power system of all would have been the nuclear-powered engines that were being developed for U.S. Air Force bombers in the 1950s. While the nuclear-powered bomber's crew would have been shielded from radiation, the craft's exhaust would have been highly toxic and radioactive; so these plans were fortunately abandoned.

Wings for Lift Only

The key difference between the basic design of an airplane and of a bird or a bee is that airplanes separate lift and thrust production. With engines to produce thrust, airplane wings can be fixed and rigid. Fixed (non-flapping) wings are rigidly attached to the fuselage, which is simpler and lighter than using a flexible connection. Not only does a fixed wing do away with the hinge itself, but it also eliminates all the actuators necessary to control and move the hinge. Moreover, because the wing itself is rigid, engineers can design it to have a precise airfoil shape, which can produce very high lift-to-drag ratios under appropriate conditions (see Chapter 2). So a fixed wing is lighter and less complex and can be very aerodynamically efficient.

An aeronautical engineer might object that some airplane wings do change shape and are not perfectly rigid and immobile. True, virtually all airplane wings have movable ailerons for roll control (Chapter 4). Most modern wings also have flaps, moveable trailing-edge control surfaces that the pilot can deflect down to increase the camber and angle of attack of the wing. This change provides more lift at lower speeds (at the cost of higher drag), allowing the plane to make slower, steeper landing approaches. Larger airliners and cargo planes have highly mechanized wings with many control surfaces—for example, complex multipanel trailing-edge flaps that spread out like Venetian blinds, leading-edge flaps and slats that droop down in front of the wing, and spoilers on top of the wing.[25] The flaps and slats help the airliner fly more slowly to make landings safer. The spoilers are flat panels that pop up on the top of the wings to kill lift after touchdown in order to keep the wheels firmly on the ground. Spoilers on some airliners also tilt up slightly in concert with the ailerons to increase aileron effect at low speeds.

All these devices are clever ways to improve the performance of large airplanes; but with the exception of ailerons, all are used in a particular setting and then ignored until a later phase of the flight. None is adjusted continuously to accommodate or produce moment-by-moment changes in the flight path. Moreover, none produces more than a fraction of the shape or area change of a bird's wing or the camber changes of a bat's or a house fly's wing during an ordinary wing stroke.

An airplane engine is one of the heaviest, densest airplane components, and where the engine or engines are attached can affect the required structural strength of the wings. Single-engine airplanes typically have piston or turboprop engines in the nose or jet engines in the tail of the fuselage. Some multi-engine jet airplanes also have the engines mounted near the tail—for instance, the common DC-9/MD-80 series of airliners as well as all business jets. Many other multi-engine airplanes have the engines mounted on the wing.

If you need more than one or two engines, wings are very handy places to hang them; and mounting engines on the wings is actually a structural advantage. If the weight of the engines is concentrated in the fuselage, it is also concentrated in the center of the wings. By moving the engines out onto the wings, however, the weight is now spread out over the wings, and the fuselage is lighter. The wing structure does not have to be as strong, so the wing can be lighter. Two or more turboprop or piston engines are limited to wing mounting because the diameter of the propeller makes hanging them on the side of the fuselage impractical. Four-engine airplanes flying with two jet engines on each wing are (and have been) quite common; they include airliners such as the Boeing 707 and 747 and the Airbus A-340 and military cargo planes such as the C-5 Galaxy and the C-17 Globemaster III. The massive Convair B-36 intercontinental bomber of the 1950s had six piston and four jet engines all mounted on or in its wings; and the slightly smaller, slightly younger, but still serving Boeing B-52 bomber has eight wing-mounted jet engines. At the other end of the scale, a few two-engine jet airliners have wing-mounted engines, including one of the most common of all airliners, the Boeing 737. Nevertheless, many smaller and medium-sized jet airliners have engines mounted on the aft fuselage, and designers have even shoehorned as many as three jet engines onto the tails of airplanes such as the Boeing 727. For these designs, advantages such as shorter (and lighter) landing gear and quieter passenger cabins outweighed the potentially lighter wing structure available for wing-mounted engines.

Pros and Cons: Animals and Airplanes As Polar Opposites

Muscle, the power source for flying animals, dictates thrust from flapping wings. The power source for airplanes, whether they have propeller or jet engines, allows the wing to be optimized for lift production. Neither approach is fundamentally better than the other, and the pros of one arrangement are largely the cons of the other arrangement.

Animals have made a virtue of necessity by integrating their highly mobile, flexible wings with their power system. Because their wings are so adjustable, their flight is also extremely adjustable. Flying animals are extremely maneuverable, much more so than even the most aerobatic airplane (see Chapter 4). Scaling plays a role here as well. At small sizes, muscle is powerful for its weight, and wing and skeleton structures can be very strong and light. Small flyers such as bees, dragonflies, hummingbirds, and finches have a great deal of power for their weight so have no problem taking off or landing vertically, which adds to their maneuverability. As animals get bigger, these advantages wane, to the point that the largest flying animals—eagles and condors—depend on soaring rather than powered, flapping flight (see Chapter 7).

Airplanes, on the other hand, do not seem to have reached an upper size limit. When the Boeing 747 airliner and the Lockheed C-5 cargo planes first flew about thirty years ago, they were considered to be gigantic. Now Airbus is beginning to produce its new super-jumbo jet, the A-380, which can carry almost twice as many passengers as a 747 can. The current constraints on airplane size have more to do with the size of airport gates and runways than with aerodynamics. A few airplane designers have claimed that even monsters such as the A-380 super-jumbo do not approach any inherent structural or aerodynamic size limits, although this appears to be a minority opinion among aerospace engineers.[26]

Airplanes can also get quite small and, as they do, tend to approach the high power-to-weight ratios of flying animals. Radio-control airplanes are perfectly conventional, functional airplanes that just happen to be too small to carry a person. Mine have wingspans ranging from twenty-five to eighty inches. Many hobbyists fly indoor models with wingspans of ten inches or less. Flying model airplanes tend to have much higher power-to-weight ratios than do their full-scale brethren; and the smallest ones tend to be absurdly overpowered, even with tiny motors. Still, even with so much excess power and a size range comparable to birds', small R/C models are more maneuverable than full-size airplanes but not quite as maneuverable as flying animals. And once we move to full-size, person-carrying airplanes, the power, inertia, structural strength, and control authority simply do not allow airplanes to approach anywhere close to the agility of a common swallow or the house fly it chases.

Artificial or Real? The Engineer's View of Lift and Drag

Since before the time of the Wright brothers, airplane designers have worked to create wings that produce lift and engines that produce thrust. Logically, engineers describe lift as an upward force to counteract weight and thrust as a forward force to overcome drag. For airplanes, this description makes perfect sense, and these concepts are firmly embedded in our aerodynamic theory and practice. But are lift and thrust really two separate and discrete forces to a flying animal? When a bird flaps its wings, it generates a net force directed up and forward. We can mathematically separate this force into an upward component and a forward component, but does the crow or dragonfly perceive them as separate? The direction of force on a flapping wing during the downstroke is mostly up near the shoulder, gradually shifting to mostly forward near the wingtip. I argue that, from a bird's perspective, trying to separate the wing's force into lift and drag is functionally irrelevant. The average direction of the overall force is what matters: tilt it forward to go faster; tilt it up to go slower. All flying animals know, either instinctively or through trial and error and lots of practice, how to adjust the direction and size of the force on the wing with great precision. If the first aerodynamics researcher had been a bird instead of a human, would we even have lift and drag as separate concepts?

~ 4 ~

To Turn or Not to Turn

I am driving the car down my street. When I get to my driveway, I want to turn in. To turn, I do something that seems so natural and obvious that it is nearly intuitive: I turn the top of the steering wheel in the direction I wish to turn the car. This angles the front wheels, which steer the car around the corner. Turning a boat with a steering-wheel helm is just as intuitive. The helmsman turns the top of the wheel toward one side, causing the rudder to angle toward that side and thus turning the boat in that direction.

The rudder, a vertical blade or wing-like device under the back of the lower hull, is the main steering device on most boats. Airplanes also have rudders located at the rear of the craft; and like a boat's, they can swing the front of the airplane to the left or right. Yet this is not the most effective way to turn an airplane in flight. Moreover, controlling turns in flight is quite different from controlling turns in boats or autos.

Bird Turns: It's All in the Wing

Bird tails do not include rudders, yet anyone who watches finches or sparrows maneuvering among tree branches knows that birds can turn quite nicely without them. If they do not use rudders, how do they manage? The secret is in the wings: both flying animals and airplanes use their wings to turn.

Imagine a crow flying over a cornfield high enough so that he can easily scan a broad area for food. The crow comes to a road perpendicular to his flight path. Down the road to his right, he spies a dead opossum, apparently killed while crossing the road. The crow wants to investigate dining possibilities, so he decides to descend and turn right. He stops flapping and glides to descend, then tilts his body so that his right wing goes down and his left wing goes up. In this tilted, or banked, attitude, the crow turns to the right.

How does he execute a banked turn without a rudder? What the crow does is to twist his wings in opposite directions. He twists the back (trailing edge) of the right wing up, and the back of the left wing down. The right wing, now at a lower angle of attack, produces less lift; the left wing, at a higher angle of attack, produces more lift. This causes the crow to tilt, or roll, to the right. When the crow is banked to the right, the total lift produced by the wings is also tilted to the right (fig. 4.1). Tilting the lift force to the right means that it tends to pull the crow to the right, and this reoriented lift actually pulls the crow around the turn. Once the crow is banked to the right, he can more or less equalize the angles of attack of the two wings (remove the twist) because, as long as he stays banked, he will automatically turn. In fact, to stop turning, he actually has to reverse what he did to roll into the bank. He twists the back of the right wing down and the back of the left wing up so that the right wing now produces more lift. The extra lift of the right wing rolls him back upright and completes the turn.[1] If he has judged correctly, he is now heading toward a delicious roadkill snack.

The central feature of an aerial turn, and one that is at odds with our everyday experience, is that the bank (tilted orientation) itself causes the turn. By banking, the bird tilts its lift to one side. That sideways tilt means that part of the force is

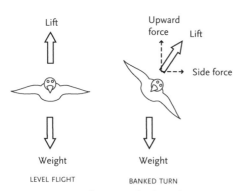

Figure 4.1 How rolling into a bank produces a turn.

directed to the side, specifically to the inside of the turn.[2] This portion of the lift to the side is what pulls the bird around the corner.

Automobiles and boats also tend to lean as they go around corners, somewhat like banking, but that tilt is essentially a byproduct of the turn. Automobiles tilt when they turn a corner because of their suspension geometry or sometimes because the road itself is tilted to counteract centrifugal force. (Most cars lean into turns, but one rare, unconventional suspension arrangement can actually cause the car to lean to the outside of turns.) Small boats with outboard motors bank into turns because the whole engine, including the propeller, swivels to produce a turn. When the propeller, located well below the boat's center of buoyancy, pushes sideways, it also tends to tilt the boat. The rudder on a small boat with an inboard engine has a similar but weaker effect. Large, fast ships, on the other hand, lean (or heel) to the outside of turns because the center of gravity of such a ship is above the center of hydrodynamic drag acting on the hull and centrifugal force acts only on the center of gravity.[3] In any case, a boat or a car does not turn because of its lean in a turn, whereas a flyer in a bank certainly turns because of that bank.

To Flap, Perchance to Turn . . .

Now we know how our crow banks when he is gliding. What if he wants to turn while flapping? Imagine he is flying along after lunch, and he sees his mate flying off to his left at the same altitude. He decides to join her, so he needs to turn left. He still turns by banking; but now, because of possible adjustments to the wing-beat pattern, he has a larger toolbox of tricks that he can use to perform the roll into the banked orientation.

For instance, he can choose to do basically what he did while gliding: tilt the trailing edge of the inside (left) wing up and the trailing edge of the outer (right) wing down constantly throughout the stroke. He can also choose to do a few other things. For example, he can make the same sort of adjustments but only at a particular point in the stroke cycle—say, during part of the downstroke. Or he can flap his right (outer) wing harder so that it moves up and down more than his left (inner) wing does; in other words, he can increase the size of his stroke on one side. If he does this, his right wing is now effectively moving faster than his left wing so his right wing will produce more lift to roll him to the left. In my own research, I have seen dragonflies essentially stop flapping with the wings to the inside so that those wings are gliding while the outside wings continue flapping. If our crow wants to turn in a hurry, he can partly flex, or fold, his left (inner) wing while flapping it less, which reduces its surface area and hence its lift. The action will quickly roll him into a very steep bank. This particular trick is probably more useful for small birds (such as finches and wrens) and bats than for our crow: the small flyers flap their wings so fast that they can afford to give up a stroke or two and not fall very far (the bird's total lift is much reduced, after all). A crow, however, might lose substantial height if he turned this way. House flies and some of their relatives take this variation to an extreme. Scientists have discovered that flies can actually fold back one wing over their abdomen (into the resting position), which produces a very abrupt turn. Again, flies can get away with this because of their very high wing-beat frequency: even if the turn lasts two or three wing beats, it is over so quickly (less than one-tenth of a second) that the fly does not have time to fall any significant distance.[4]

Some insects have more turn tricks than birds and bats do. Most flying insects have two pairs of wings, although they are often flapped together so that they act as a single pair. A few insects use their front and hind wings quite independently, which gives them great maneuverability. Dragonflies are the premier example. They can use their front wings, their hind wings, or both to roll into a bank; and they can use many possible combinations of wing twisting and increasing or reducing the size of the wing stroke. Not surprisingly, dragonflies are among the most maneuverable of flying animals.[5]

Remember that, for all their variety and complexity, these wing adjustments are just for rolling the animal into a bank. In other words, these tricks really just initiate the turn. Once in the bank, the flyer can return to a more or less equal pattern on both sides, and the banked attitude is what actually turns the animal. And of course, at the end of the turn, the animal must reverse those adjustments to roll upright and stop turning.

Hovering

Regardless of which tricks they use, all flying animals in typical forward flight turn using the banked-turn process. The only significant exception is during hovering and very slow forward flight. The banked-turn process does not produce a normal turn without some forward speed. If a hovering hummingbird adjusts her wing beat to tilt her body to the left, the little bird simply flies sideways to the left without actually turning. If she wants to actually turn to the left, she does something more complex and subtle (and, frankly, not entirely understood) to pivot and face left while hovering. Physically, we know that the hummingbird must adjust her wing strokes so that the right and left wings produce different amounts of drag while still producing equal amounts of lift, but scientists do not yet know exactly how she does this.

Airplanes Are Not Boats

The early pioneers of heavier-than-air flying assumed that they would turn their craft with rudders, just as they would turn a boat. At least a couple of these inventors built small-scale models that flew quite nicely. Experimenters such as George Cayley, Alphonse Penaud, and Samuel Langley did not see, however, that the very stability traits that allowed their scale models to fly so well made their full-scale designs very difficult to turn.[6] Depending on a rudder to turn the craft simply made matters worse. A rudder can turn an airplane, but it produces a sluggish, skidding turn with a variety of disadvantages, as we will see later in this chapter. If the early designers had actually gotten their contraptions to leave the ground, their rudders may have lost the battle against stability in the fight to turn their machines.

The Wright Brothers' Huge Insights

Wilbur and Orville Wright were the first to fly successfully largely because they were the first to figure out how to turn. They had watched soaring vultures and had noticed how these big birds righted themselves by twisting their wings when accidentally banked by a gust. They concluded that birds used this same trick to bank intentionally, and they realized that banking was the most effective way to turn. They concluded that they would need to have some means of roll control (that is, of tilting the craft right or left) and that this mechanism would be their primary turn control.[7] They chose to emulate the mechanism used by birds.

Popular myth portrays the Wrights as unlettered mechanics who invented the airplane by clever tinkering. This tale ignores their enormous intellectual and scientific achievements. In fact, they approached their task methodically, precisely, and very scientifically. They read all they could on aeronautics, including highly technical works by Otto Lilienthal and other researchers, and they struck up a lively and cordial correspondence with Octave Chanute, perhaps the foremost American engineer of the time. At first, they used data from previous researchers, particularly Lilienthal, as the basis for their designs.[8] Eventually they began to carry out their own research to correct and extend the research of their predecessors. Their propellers, for example, were so efficient that at least a decade passed before anyone else achieved propellers of equal performance.[9] Once they actually started designing their machines, the Wrights tested them in carefully controlled stages. They started with a large controllable kite to test their roll control mechanism, which they called wing warping. This mechanism was basically the same twisting process used by gliding birds: the Wrights raised the trailing edge (lowered the angle of attack) of one wing while simultaneously lowering the trailing edge of the other wing. They planned to use wing warping to level the wings from unintended banks and to intentionally bank them to produce turns.[10]

In contrast to earlier experimenters (and more than a few later ones), the Wrights explicitly designed their machines to be unstable; pilot skill was needed to keep them aloft.[11] They thus had to teach themselves to fly. After using their kite to convince themselves that wing-warping worked, they built a glider and looked for a treeless test area with strong, reliable winds and low hills. They chose Kitty Hawk, on the Outer Banks of North Carolina.

The first glider they brought to Kitty Hawk in 1900 was essentially a large set of biplane wings with a small pivoting wing suspended in front to act as an elevator (fig. 4.2). The glider could not produce enough lift to carry a person except in very high winds, so they mostly flew it as a kite to explore their control systems. The next year they brought a larger glider to Kitty Hawk, which they calculated should easily carry a person. Although they made many successful glides, they were disappointed with the 1901 glider for two reasons. First, it did not produce nearly as much lift as they had calculated it should, based on Lilienthal's tables. (The Wrights did not take into account the differences in aspect ratio or airfoil shape between their glider and Lilienthal's wing model.)[12] Second, sometimes they could turn with wing warping, but other times they ran into an unexpected problem. Occasionally when they banked the glider, the high wing seemed

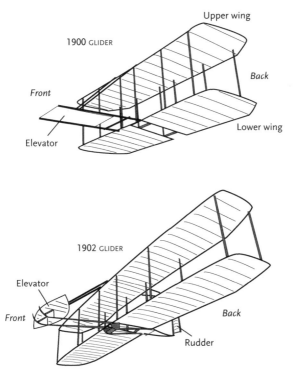

Figure 4.2 The Wright brothers' 1900 and 1902 gliders in three-quarter view from below.

to pull back, dragging the front of the craft in the wrong direction and causing it to slide sideways into the ground. (How much longer would it have taken them to perfect their airplane without the soft Outer Banks sand dunes to crash-land on and reduce damage to machine and pilot?) Another problem occurred when the glider was already in a bank: sometimes when they applied wing warping in the opposite direction to right the craft, the low wing would pull back and send the glider spinning to the ground. Orville and Wilbur called this behavior "well digging," and they were puzzled and concerned by it.[13] They embarked on an innovative research program to develop their own aerodynamic tables and design better wing shapes. Their next glider benefited greatly from this new information.

The third (1902) glider had longer, narrower wings with more wing area and a pair of flat, immovable, rudder-like vertical surfaces mounted on struts behind the wings (fig. 4.2). The brothers hoped that these vertical surfaces would act as a weather vane and keep the glider from sliding sideways when banked. This glider was indeed a great improvement over the previous one, but it still occasionally displayed annoying well-digging behaviors.

In modern aeronautics, we would call their problem *adverse yaw,* borrowing the sailing term *yaw* to describe the way in which the front of the airplane swings to the right or the left. Adverse yaw comes directly from wing warping. When the pilot twists the wings for a right bank, the left (outside) wing not only produces higher lift but also produces higher drag. This higher drag tends to swing the glider's nose to the left, even though it is in a right bank. Regarding the Wright Flyer, whether or not adverse yaw became strong enough to be a problem depended on circumstances. If the controls were mostly neutralized once in a bank, little or no adverse yaw occurred. In other conditions, particular combinations of airspeed, wind gusts, and shifts in center of gravity might have forced the pilot to hold wing warping throughout the turn; and in such cases, the dreaded well digging was likely to take place.

In the middle of the 1902 gliding season, the Wrights had a crucial insight that paved the way to successful powered flight: if they could turn the rudder in the direction of the turn while they warped the wings, they could overcome adverse yaw. They replaced the fixed vertical fins with a movable rudder and wired it into the wing-warping system so that the rudder would automatically turn when the wings were warped.[14] The Wrights went on to make more than 1,000 glides with this glider, which was the machine on which they actually taught themselves to fly. They made glides of more than six hundred feet and flew in winds of up to thirty-five miles per hour. By the end of that season at Kitty Hawk, Orville and Wilbur probably had more flying experience between them than all other living experimenters combined. (They calculated that, by the time he died, Lilienthal had about five hours of total flight time spread over about five years. Based on the number of glides they made and their typical distances, I estimate that the Wrights achieved between three and four hours of flight time during the 1902 season alone.)[15] Given that the Wrights solved the control problem and taught themselves to fly on the 1902 glider, some historians think that the machine is more important than their first powered craft.[16] At this point, practical powered flight was largely a matter of finding an appropriate engine, designing propellers, and scaling up the glider to carry the additional weight.

As I see it, the Wright brothers had three unique insights, all related to turning, that were prerequisites for designing practical airplanes. First, they decided that the pilot would have to actively fly the craft in three dimensions, not simply steer it around like a wagon. Second, they realized that using the wings to bank was the most effective way to turn (while also contributing to three-dimensional controllability). Third, they realized that wing warping was not the complete answer to turning and that they needed a movable rudder to overcome adverse yaw. Airplanes today still use banked turns (although we will see that wing warping has been replaced by other devices), and pilots still use their rudders to counter adverse yaw (see Chapter 5).

Enter the Aileron

The Wrights' biplane design, with its wire-braced, springy, boxy biplane wings, was ideal for the bird-like wing-warping system.

As long as airplanes had flexible wooden wings stiffened by wire trusses and carried relatively light loads, wing warping was a practical method of banking. But as airplanes flew faster and became heavier, the flexible, springy wings needed for wing warping became impractical. Moreover, some early airplane builders were concerned about the complexity of the wing-warping wires and their potential for stretching and breaking in flight, with potentially disastrous results. Several designers experimented with movable auxiliary surfaces to replace twisting of the whole wing. These devices soon became known as ailerons and gave rise to the hinged, flap-like surfaces called ailerons that are still used to turn most airplanes today.

Some confusion surrounds the derivation of the term *aileron.* Although the word is the diminutive of the French *aile,* or wing, it is also the literal translation of "pinion," or a bird's primary feathers. According to eminent aviation historian Charles Gibbs-Smith, in nineteenth-century French, *aileron* was also used in a variety of contexts to mean "winglet."[17] Although most English-language references describe the term as French for "little wing," that meaning is not common in more recent French, where it just means "pinion."

The earliest ailerons were separate plates or miniature wings located in a variety of places on various craft. Robert Esnault-Pelterie's 1904 glider had them in front of the wings, Louis Bleriot's unsuccessful 1906 No. IV had them behind the wings, and Glen Curtiss's 1909 Biplane and S. F. Cody's 1908 British Army Aeroplane No. 1 both had them between biplane wings. Closer to the modern arrangement, Curtiss's June Bug and Bleriot's No. VIII, both from 1908, had them extending from the wingtip.[18] The modern configuration of a hinged flap on the trailing edge of the wing was actually used on several unsuccessful designs before Henri Farman adopted it on his successful airplanes of 1908 and 1909. The structural simplicity and aerodynamic effectiveness of Farman's ailerons soon led most other designers to adopt them, and they have been the standard roll-control device on airplanes ever since.[19] Curiously and poignantly, the Wright brothers, who had been so innovative and insightful up until 1905, continued to use wing warping on their airplanes until 1915, long after most others had switched to ailerons.

One last roll-control device is worth mentioning. The spoiler is a panel on the top of a wing that rises up and "spoils" the smooth flow of air over the wing. If a spoiler can be used on one side at a time, it can be a very effective roll-control device, with the added advantage of creating no adverse yaw. The spoiler is raised only on the wing to the inside of the turn, not on the outside wing; and because it increases drag as well as decreases lift, it actually helps swing the nose into the turn. Birds could, in principle, use spoilers as ailerons if they could independently raise a patch of feathers on the top of one wing, but no one has ever seen birds do this. Presumably, their great ability to change their wings' shape removes any need for spoilers. One disadvantage of spoiler ailerons is that they can be much easier to deflect than conventional ailerons, so the pilot does not get much tactile feedback and can inadvertently overstress the airplane. Nevertheless, some World War II military airplanes, including the P-61 Black Widow night fighter, used spoilers for ailerons, and several modern airliners use them in combination with conventional ailerons.[20] As different as spoilers are from wing warping or conventional ailerons, they still work by banking the airplane to produce a turn.

What about Rudders?

We have seen that the rudder is important for overcoming adverse yaw. Properly used, it keeps the nose pointing into the turn, counteracting the ailerons' tendency to swing the nose to the outside and cause a slip. Pilots call this use of rudder "coordinating a turn," and it is the most important function of the rudder in routine flight. So while the rudder does play a role in turns, that role is a minor and supporting one. Most airplanes can actually turn with ailerons alone, although at less than peak efficiency.

Pilots also use rudders in other situations, such as when they are keeping the nose aligned with the runway during take-offs and cross-wind landings. Sometimes pilots even need to sideslip the airplane intentionally—for example, to lose altitude without losing airspeed. In Chapter 5, we will look at these uses for the rudder in more detail.

Okay, You Can Turn with a Rudder

At this point, most readers who have experience in flying radio-controlled models are saying, "Wait a minute! I have a flying model with no ailerons that turns just fine using the rudder." In fact, quite a few flying model airplanes are intentionally designed to turn solely using the rudder. Why have designers chosen to build model airplanes in a way that seems to run counter to most of what we have learned in this chapter? Why do they use rudders for turns?

No one has yet written an authoritative history of model aviation, but I suspect the answer involves the evolution of free-flight models into R/C models. Flying models predate radio systems available to hobbyists by two or three decades. Rubber band–powered, stick-and-tissue models have certainly been around since the 1920s, perhaps even earlier. By the 1930s, people were building larger, more sophisticated models with tiny, temperamental, unreliable gasoline engines. These were all models that we would now call free-flight models because, once in the air, they fly without any control from the ground. The idea is that, when properly adjusted, the model flies itself. The models were adjusted to climb when the engine was running, glide (descend) when the engine stopped after running out of fuel, and usually to make a shallow turn so as to fly in wide circles. The builder typically tossed the model into unpowered glides a few times before making a powered flight. These test glides gave the builder a chance to make any adjustments needed to keep the model in stable flight. Obviously, stability is at a premium in free flight because the model must be able to fly without any input from a pilot.

One of the more effective ways to stabilize any flyer is to give its wings dihedral. Dihedral means that each wingtip is elevated slightly above the root (where the wing attaches to the fuselage). Wings with dihedral form a shallow V when viewed from the front (fig. 4.3). Dihedral stabilizes flight by effectively canceling the effects of small roll or yaw disturbances. An unintended roll induces a bit of sideslip toward the side of the low wing, which slightly increases the low wing's angle of attack and causes it to roll back upright.[21]

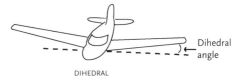

DIHEDRAL

Perhaps employing the principal of "If a little is good, a lot must be better," modelers discovered that they could get even more stability if their wings had double dihedral, or polyhedral. Wings with double dihedral have a sec-

POLYHEDRAL

Figure 4.3 Wings with dihedral and polyhedral.

ond shallow V-angle added between the tip and the root of each wing, a design that provides even stronger stability (fig. 4.3).[22] By the 1950s, free-flight models had become highly developed: simple, reliable glow engines replaced the much-cursed gasoline engines, and some of the larger models had wingspans of well over five feet. Flight times of both rubber band–powered and engine-powered models were measured in minutes rather than seconds.

During World War II, the military tried using large R/C models as target drones for gunnery training, but the control equipment was cumbersome, extremely expensive, and too heavy and bulky for hobbyist use.[23] By the mid-1950s and early 1960s, modelers were experimenting with homebuilt or kit-based R/C systems. Naturally, they started out by adapting free-flight models. Not only were such models readily available and familiar, but their designs gave them some hope of avoiding catastrophic crashes in the event of all-too-frequent radio failures. Most of the earliest R/C systems could only operate one flight control on the model, and modelers almost always chose to control the rudder with these systems. Modelers would trim their planes to climb with the engine on and glide with it off, as before. Instead of flying in a fixed circle, however, they could (they hoped) steer the model right or left with the rudder.

Why did these modelers choose rudders instead of ailerons for turn control? Several factors played a role. Some modelers may have simply had the mistaken impression that full-sized airplanes used rudders to

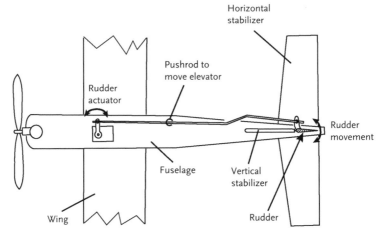

Figure 4.4 R/C model rudder, showing linkage to its actuator (top view).

turn. But I am inclined to think that this was not a major factor because serious airplane enthusiasts (particularly those willing to spend the large amounts of time and money required by early R/C models) would have known about ailerons. Two other factors were probably more important. First, the mechanical linkage for ailerons is more complex: two surfaces must be moved in opposite directions as opposed to a single rudder. Double dihedral makes things worse. If a mechanical actuator is in the fuselage (as required by the bulky early systems), mechanical linkages through the outer dihedral angle to ailerons near the wingtip become complex and inefficient. In contrast, the mechanical linkage for moving a rudder couldn't be simpler: a single, straight rod from the actuator in the fuselage to a control horn on the rudder (fig. 4.4).

The other important factor was the aerodynamic behavior of free-flight models. Craft whose wings had large dihedral angles, or double dihedral, were so stable that ailerons tended to be relatively ineffective. Conversely, the pronounced dihedral (or polyhedral) of these wings enhanced the rudder's turning ability. Deflecting the rudder yaws the model, swinging the nose to one side. For example, deflecting the rudder for a left turn swings the model's nose to the left, but the craft sideslips, or briefly continues to fly straight. As it sideslips, the right wing swings around toward the front. The wing is angled up because of the dihedral. Visualize this as a situation in which the oncoming wind strikes the bottom of the wing and pushes it up, although what really happens is that the right wing's angle of attack increases slightly, increasing its lift. This raises the right wing and tilts the model into a left bank. In other words, if a model with substantial dihedral is intentionally yawed to one side by its rudder, it also tends to roll to that side. So the rudder can, and does, cause such a model to roll into a bank.

A properly designed R/C model can thus use a rudder to turn effectively. Turns using rudders are never as crisp or precise as turns using ailerons; but for slow, relatively stable flying models, rudder-only turns are entirely adequate.[24] These are the characteristics of many R/C trainer models still being sold today. Due, however, to the enormous increase in reliability and decrease in size (and cost) of modern radio systems, the vast majority of modern models use the same three primary flight controls—aileron, elevator, and rudder— plus engine throttle as full-sized airplanes do. Some even feature extra controls for niceties such as retractable landing gear or flaps or for dropping toy parachutists or simulated bombs.

Even So, Full-Sized Airplanes Have Ailerons

To the best of my knowledge, aside from a few pioneering airplanes built before 1911, no successful production airplanes have flown using only the rudder to turn. Why not, given its success on models? For one thing, giving up ailerons means giving up the ability to level the wings quickly and precisely, exactly the use originally envisioned by the Wright brothers for wing warping. Imagine landing in a light airplane and being tilted to one side by a gust. The pilot's ability to level the wings quickly may be a life-or-death matter. Although the rudder can be used to level the wings in some airplanes, it also tends to change the airplane's heading (definitely undesirable when landing) and may trigger a spin if used too vigorously.

One significant difference between full-sized airplanes and flying models is that full-sized airplanes are not usually built with anything approaching the strong stability of a free-flight or early R/C model.

In fact, the Wrights saw stability as a menace because they felt that too much stability prevented the pilot from flying out of dangerous situations (and, they believed, led to the deaths of some aviation pioneers). By modern standards, they went to the opposite extreme, putting the elevator in front and giving their wings negative dihedral so their airplanes were quite unstable.[25] Modern airplanes are the result of a compromise: they are usually stable enough to fly hands-off for short periods but unstable enough so that the pilot can easily change directions when he or she wants to. After all, the pilot's function is to control the airplane; so full-sized airplanes cannot be as stable as free-flight models, and rudders will not be effective for controlling turns.

How Not to Turn: Stability

What if a flyer does not want to turn? Suppose an airplane pilot wants to fly in a straight line. Naïvely, we might assume that flying in a straight line would simply be a matter of not turning. In the imperfect real world, however, things are more complicated.

The ease or difficulty of flying a straight course depends on a flyer's stability. Stability is the tendency of a flyer to return to the original course after some disturbance—a gust, say, or an accidental control movement.[26] Highly stable flyers have a strong tendency to return to their original course, whereas a small disturbance can throw an unstable flyer off its course. Free-flight model airplanes are extremely stable, full-sized trainers and large transport airplanes are fairly stable, birds and fighter airplanes are not very stable, and insects are very unstable. Why such variations in stability?

Stability is a two-edged sword. High stability means a straighter flight path and less work for the pilot when an airplane is flying straight and level (what most airplanes do most of the time). The drawback is that a very stable airplane responds slowly to the pilot's controls, in effect resisting his or her attempts to change direction. If maneuverability is important, high stability is often a hindrance.

After the Wright brothers, as airplanes became more practical, designers attempted to build various amounts of stability into them. Because a stable airplane is easier to fly, most designers throughout the twentieth century tried to build in as much stability as possible, while keeping in mind the maneuvering requirements of a given airplane's intended use. Many factors affect an airplane's stability, including the dihedral angle of the wings (which works just as well on full-sized as on model airplanes), size of the tail and the tail's distance from the wings, and whether the wings are on top of the fuselage (high wings, more stable) or on the bottom (low wings, less stable).[27]

A stable flyer, whether engineered or biological, requires less mental effort to fly. The pilot or flying animal does not need to respond actively to every little disturbance. The inherent stability of the design takes care of most of the minor corrections. Training airplanes tend to be stable for this reason. Similarly, biologists expect the most primitive flying animals to be relatively stable because they have not yet evolved the specialized nervous system needed by an unstable, highly maneuverable flyer. Both the earliest pterosaurs and Archaeopteryx (the earliest known bird) had very long tails, which make powerful stabilizers, whereas later pterosaurs and modern birds and bats have very reduced tails, bearing out this prediction.[28]

Stability, however, is not always a good thing. Most flying animals are extremely maneuverable. When catching prey or avoiding a predator is a life-or-death matter, high maneuverability is a distinct advantage. Stability degrades maneuverability, so most flying animals are more or less unstable. This quality allows the animal to make sudden, sharp turns but also means that flying in a straight line requires some effort. To replace the built-in stability of shape, these animals use active stabilizing mechanisms, or what engineers call stability augmentation. Flying animals quickly sense unintended course deviations and make frequent, rapid control movements to correct them. Animals can use a variety of sensory cues to detect course deviations. Birds and most insects depend on vision and sensing air flows over their bodies. Bats also use echolocation (sonar), and a few insects even have a gyroscopic heading sensor (see Chapter 6). These senses trigger reflexes that drive a complex set of control responses, which are similar to normal control movements (wing-beat adjustments) but are quicker and smaller. These reflexes are built into the animals' nervous systems, the result of natural selection acting to favor animals with high maneuverability.

In a curious convergence, the Wright brothers deliberately designed their first airplanes to be easy to maneuver, which made them unstable.[29] They did not do this in imitation of birds but because they thought stable flying machines were dangerous. As we have seen, later designers found that they could increase an airplane's stability to make it easier to fly without making it less safe.

Fighter airplanes are the major class of airplanes that have generally favored maneuverability over stability (see Chapter 9). As for flying animals, maneuverability can have life-or-death consequences in an aerial dogfight. Airplanes with mechanical controls (the pilot's stick directly connected to the control surfaces by cables or rods) require some built-in stability or they become unflyable. Designers of World War II fighters, for instance, tried to make them as maneuverable as possible; but those planes still retained enough stability to be safe and comfortable to fly when not in combat. Advances in computer technology have changed this picture. Many modern fighter airplanes have computerized systems that can detect course deviations and correct them automatically. In these automatically stabilized "fly-by-wire" systems, the pilot's controls are connected to a computer rather than to the control surfaces.[30] When the pilot moves the control stick, the computer interprets the command and decides how to move the control surfaces to carry out the desired maneuver. The computer then sends electrical signals in wires to motors or hydraulic cylinders that actually move the surfaces. In addition to these intended course changes, the computer also detects and automatically counters any unintended course changes. The computer can be programmed to provide as much or as little apparent stability as desired, so pilots can fly airplanes that would be too unstable to fly without the computer.[31] (Actually, fly-by-wire airplanes usually have three computers operating under majority rule for reasons of redundancy and fault tolerance. Moreover, fly-by-wire is lighter, more redundant, and more damage-resistant than mechanical controls; so most new jet airliners now have fly-by-wire controls.) Such unstable fighters can be extremely maneuverable, and most frontline fighter jets—the U.S. Navy's F/A-18 Hornet, the U.S. Air Force's F-22 Raptor and F-117 Stealth Fighter—use this approach (discussed in Chapter 9), which brings us full circle back to the Wright brothers and their views on stability and control.

In the final analysis, Orville and Wilbur got the control concept right. We still turn airplanes by rolling into a bank, although we do it with ailerons rather than by warping the wings. We still use the rudder mainly to overcome adverse yaw, not as a primary turn control. The Wrights' insight about the necessity of controlling a flying machine in all three dimensions was a key factor in the development of flying machines. Their views on stability were not adopted by later airplane designers, but we still have to nod in their direction. Designers now design deliberately unstable fighter airplanes, not for safer flight but for increased agility. Even so, modern fighters share their lack of stability with the very first successful machine to fly.

~ 5 ~

A Tale of Two Tails

At her back end, a goose has a horizontal fan of feathers that she can spread out or fold up. At a Boeing 747 airliner's back end, there are two surfaces: a horizontal one that looks like a small wing and a vertical one that looks like half of a small wing. We call both of these structures *tails*. Although *tail* is a perfectly sensible term for "whatever is on the back end of some creature or object," it can also be misleading because it obscures the huge structural and functional differences between the tails of flying machines and those of avian flyers. Using the same name for these two structures would be rather like using the same name for automobile tires and the paddlewheel blades on a nineteenth-century steamboat. True, both structures go around and push the vehicle forward, but they operate in different ways.

Different Structures

The most obvious difference between airplane tails and bird tails is that airplanes (with very few exceptions) have a vertical tail surface and birds do not. This part of the airplane's tail is called the vertical stabilizer, and the movable trailing-edge section is called the rudder. As I explained in Chapter 4, an airplane's rudder, unlike a boat's, is not a pilot's primary control for turning; pilots generally use ailerons instead. An airplane's rudder can swing its nose to one side, but mostly this causes sideslip, which is a very inefficient way to turn. During a normal banked turn produced by ailerons, the rudder is actually more useful for preventing sideslips due to adverse yaw.

An airplane's horizontal tail surface is called the horizontal stabilizer, and its hinged trailing-edge control surfaces are called elevators. Whereas the vertical stabilizer only extends up from the fuselage (making it look like half of a wing), the horizontal stabilizer extends out to both sides of the fuselage. In the most common arrangement (the conventional tail), the horizontal stabilizer is located at the very back of the fuselage; but it can also be partway up the vertical stabilizer (a cruciform tail), on top of the vertical stabilizer (a T-tail), or even on the fuselage in front of the main wing (a canard stabilizer). The T-tail has become common on airliners and business jets because a number of designs have jet engines mounted on the rear of the fuselage (see Chapter 3).

A bird's tail is strictly a horizontal surface that is effectively all feathers. At the back end of its body, a bird has a very short, stubby, triangular pygostyle—the near-vestigial, flesh-and-blood part of the tail—which carries a horizontal row of feather shafts. (The "parson's nose" of a cooked turkey is the turkey's pygostyle.) The bird can fan out its tail feathers to form a large surface or collapse them into a compact, narrow stack about as wide as a pair of feathers side by side. Tail feathers are typically about the same size and strength as primary wing feathers; but unlike the primaries, they tend to be symmetrical around the central shaft. Different species, however, show a huge range of size and elaboration of tails.[1] Some, such as grouse and ducks, have short tail feathers. Many, such as typical songbirds, have tails that are between one-third and one-half the length of a wing. A few, such as swallows and frigatebirds, have long, deeply forked tails with outer tail feathers almost as long as a wing.

The longest and most elaborate tails usually do not have much aerodynamic function. For example, the tails of male pheasants, lyrebirds, turkeys, and peacocks evolved primarily to attract females and actually impair flight efficiency. Even small birds, such as scissor-tailed flycatchers, may have elongated tails (again, mainly for sexual display) with feathers significantly longer than the rest of the bird's body. Birds have been using tail feathers to attract mates practically as long as they have been birds. Paleontologists have recently discovered some fascinating bird fossils from more than 140 million years ago, very early in the evolutionary lineage of birds. One species, *Confuciusornis sanctus*, included individuals with typical tails but also some with hugely elongated, scissor-like tail feathers. Most scientists think the long-tailed birds were males who used their long tail feathers to attract females, just as scissor-tailed flycatchers and pheasants do today.[2]

Some birds use their tails for mechanical, rather than amorous, functions separate from flight functions. Chimney swifts and woodpeckers have short, stiff tails that they use as props for perching on vertical surfaces. Woodpeckers use their tails to brace themselves on tree trunks and thus gain more leverage when whacking their beaks into trees. Chimney swifts have shorter tails, and the feather shafts are stout and extend beyond the feather blades as short, sharp spikes. These birds prop their tails against the inner surface of hollow trees or the masonry of chimneys and use them to support a large share of their weight when perching on vertical surfaces.[3]

Birds move their tails quite differently from the way in which airplanes move their tails and rudders. As mentioned, a bird can spread its tail into a broad fan, collapse it into a narrow stack, or vary its width between those two extremes. A bird can also tilt the tail trailing-edge-up or -down or twist the tail by pushing feathers down on one side and pulling them up on the other. Although tilting the whole tail up or down looks like it might have a purpose similar to the function of an airplane's elevator, bird tails actually work quite differently: spreading or collapsing the tail is usually more important than tilting or twisting it.

How Tails Work

Even without any rudder or elevator movement, an airplane's tail is important for stabilizing an airplane's flight. Visually, tails appear to be similar to the feathers on an arrow or a weathervane. Actually, however, they function like wings. When the airplane sideslips, the sideslip angle increases the vertical tail's angle of attack, thus increasing the lift on that side, pulling the tail around, and straightening the airplane. Figure 5.1 illustrates this effect for the vertical stabilizer, but the horizontal stabilizer does exactly the same thing, stabilizing up-and-down perturbations instead of right-left ones. The size and aspect ratios of the horizontal and vertical stabilizers have significant effects on the stability of an airplane and are important design considerations.

Airplane rudders are connected to a pair of pedals that the pilot operates with his or her feet. In *Airplane Stability and Control: A History of the Technologies That Made Aviation Possible*, Malcolm Abzug and Eugene Larrabee point out that, compared to most steering devices, standard rudder pedals seem to work backward.[4] To turn a bicycle or a sled, you move the handlebars or the steering bar so the front of the bar faces into the turn, which is a natural and intuitive motion. For example, to turn to the right, you push on the *left* handle. Early airplanes had a similar foot-operated rudder bar, invented by Louis Blériot, but it worked in an opposite way: to yaw to the right,

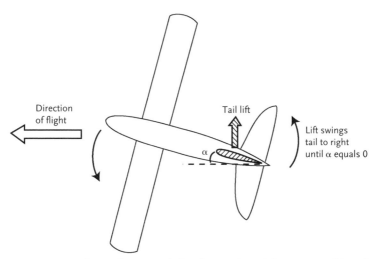

Direction of flight

Tail lift

Lift swings tail to right until α equals 0

α

Figure 5.1 Top view of an airplane in a sideslip, showing the stabilizing action of the tail. The sideslip puts the vertical stabilizer at an angle of attack (α), so lift on the tail points to the side and corrects the sideslip. In straight flight, the angle of attack is zero, so there is no sideways lift on the tail.

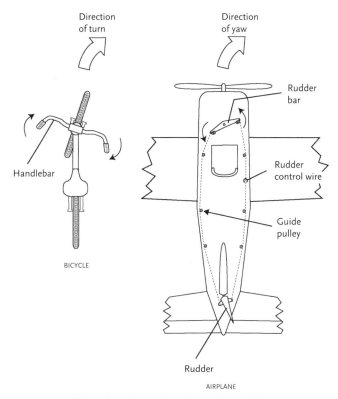

Figure 5.2 Top views of a bicycle and an early airplane, showing the opposite turn action of bicycle handlebars and the rudder bar of an airplane. Note that making the rudder bar work like the handlebars would require a more complex control-wire arrangement.

you pushed forward on the *right* side of the rudder bar. This arrangement simplified the arrangement of control wires (fig. 5.2) and became the standard convention for rudder controls. Modern airplanes have a pair of pedals to operate the rudder, but they are linked to each other and to the rudder so that they still work just like the early rudder bars. Some pilots even still call rudder pedals "rudder bars." (This mechanism may explain why, when I took flying lessons, the rudder action seemed unsettling and nonintuitive in contrast to the natural action of the other controls.) Abzug and Larrabee note that Igor Sikorsky, famed airplane and helicopter designer, so detested the conventional rudder action that he had his personal airplane rigged with reversed rudder controls. Of course, no one but Sikorsky could then fly his airplane safely.[5]

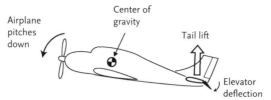

Figure 5.3 Deflecting the elevator down increases lift on the tail. This lift rotates the nose down. (Such vertical rotations are called *pitching*.) The checkered circle marks the center of gravity around which the airplane rotates.

Elevators (the movable control surfaces on the horizontal stabilizer) are controlled by either a control stick (a joystick) or a control wheel. When the pilot pulls back the stick or wheel, the elevator deflects upward. This gives the horizontal stabilizer negative camber, producing a downward force on the tail, which tends to raise the nose (fig. 5.3). A forward push on the stick or wheel deflects the elevator down, which produces an upward force on the tail and lowers the nose. This operation feels much more intuitive than controlling the rudder pedals does: pulling back on the stick raises the nose, as if an invisible string from the top of the control stick were literally pulling it up.[6]

The elevator raises and lowers the airplane's nose, so it might be perfectly reasonable to assume that this action is what makes the airplane go up and down (especially given the name *elevator*). Raising the nose both tilts up the engine thrust and increases the wing's angle of attack, which certainly cause an airplane to climb (as long as it does not stall). In fact, however, the elevator's primary use is to control the airplane's airspeed. How can this be?

Raising the nose of an airplane with the elevator increases the wing's angle of attack, which causes the airplane to climb *if* the thrust stays constant. If the thrust is reduced correctly, the airplane will fly more slowly; so at a lower speed and a higher angle of attack, the lift will be the same and the airplane will fly level. Pilots learn the appropriate power settings—revolutions per minute, manifold pressure, or percent power, depending on the engine—and air speeds for different flight maneuvers and conditions. So to climb, the pilot sets the throttle to give standard climb power and adjusts the elevator to give the airspeed for a standard rate of climb. When she reaches cruising altitude, the pilot reduces the throttle to cruise thrust and adjusts the elevator to give cruise airspeed and voilá—level flight. If she raises the nose with the elevator at cruise power, the airplane may climb, but it will also slow down so will probably climb slowly and inefficiently.[7]

If pilots use the elevator to control airspeed, what do they use to control climbs and descents? In normal (non-aerobatic) flight, they use the throttle: increased power to climb, moderate power for level cruise, and decreased power to descend. They adjust the elevator to set the airspeed, usually to give speeds for best fuel economy but sometimes to slow down for turbulence or to speed up to reduce travel time. One of the first, and sometimes hardest, jobs of a flight instructor is to train new student pilots to think of the elevator as a speed control rather than a vertical motion control. Aerobatic maneuvers are a special case, not generally relevant to most airplanes. In aerobatics, the elevator *is* the primary control for producing more extreme climbs and dives.

The rudder's function is something of a paradox. Just as the elevator raises and lowers the airplane's nose, the rudder yaws the airplane, or swings its nose left or right. In Chapter 4, we saw that the rudder was important for overcoming adverse yaw when the ailerons roll the airplane into a banked turn. As a pilot would say, the rudder helps to "coordinate" the turn. Many airplanes will turn without a rudder when banked; but in most cases, the airplane sideslips during part or all of the turn. In other words, the airplane's nose points more or less to the outside of the turn as it goes around. This is called a slipping turn. A sideslip occurs whenever the airplane's nose is not directly aligned with the direction in which it is flying, a situation that can happen during slipping turns. The airplane's drag goes up dramatically during a sideslip, which, if uncorrected, causes the airplane to slow down and descend. Coordinated turns are clearly more efficient.

Not-So-Adverse Yaw

Although pilots consider slipping turns to be a Bad Thing, sometimes a deliberate sideslip comes in handy. If the pilot concludes he may land too far down the runway during a landing approach, a sideslip can be a useful way to temporarily steepen the descent.[8] This trick is especially useful in simpler or older-style airplanes without wing flaps. Also, in some airplanes with large nose-mounted engines, the engine blocks the pilot's view down and forward. During the landing approach, when the view down and forward is crucial, a quick sideslip can literally swing the nose out of the way. Then the pilot can look down through the lower corner of the windshield or the front of the side window and get a clear view of the runway.

Rudders are important on propeller-driven aircraft for another reason. When the pilot of an airplane on the runway advances the throttle to take off, the plane has a tendency to yaw to one side as the propeller speeds up. Airplanes whose propellers turn clockwise (as seen from the cockpit) yaw to the left; the opposite is true for airplanes with counterclockwise-turning propellers. Although this take-off swing is commonly called the torque effect, several processes actually contribute to it, torque probably being the least important. For instance, the swirling slipstream produced by the propeller tends to push on one side of the vertical tail. Moreover, if the propeller's rotation is tilted from the vertical, the propeller itself produces a yawing force: the propeller blades will be at a higher angle of attack on the down-going side of their rotation (the righthand side for clockwise rotation) than on the up-going side, so they take a bigger bite of air and produce more thrust on the down-going side. This trait is especially pronounced on airplanes with tail wheels; but at some point all airplanes usually have to rotate, or tilt nose-up, to take off, so it can affect nose-wheel airplanes as well. Finally, the torque of a clockwise-rotating propeller tends to give a slight counterclockwise push to the airplane. Faced with all these turning tendencies, pilots of propeller-driven airplanes must often make liberal use of the rudder during the take-off roll to keep the airplane on the runway.[9] Jet-powered aircraft generally do not yaw much during the take-off roll. This in itself shows how little torque affects the take-off swing, given that jet engines rotate much faster than piston engines or propellers do.

Rudders are also vital for crosswind landings. Normally, pilots prefer to land into the wind, but winds do not always cooperate and blow right down the runway. They often blow at some angle to the runway. If a pilot attempts to land in a crosswind by flying as he would with a direct headwind, he will find himself being blown off to one side of the runway. He has a couple of choices for staying lined up with the runway during the landing approach, and both require careful rudder use. The easiest one to visualize is the crabbed approach. This technique involves turning the airplane to face slightly into the crosswind on the approach. The airplane's path on the ground is aligned with the runway, but the craft faces at a slight angle from the runway to compensate for the crosswind (fig. 5.4). This is just like the classic "boat crossing a river" physics problem: to go straight across the river, the boat must aim a bit upstream, just as the pilot landing in a crosswind must aim a bit upwind of the runway. Immediately before touchdown, the pilot must use the rudder to swing the nose around to line up with the runway. This last maneuver requires skill and fine

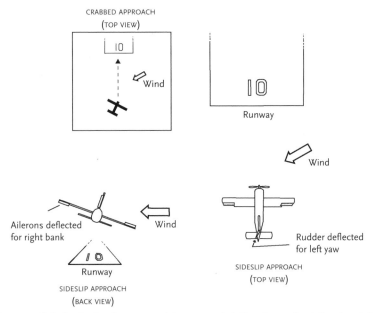

Figure 5.4 Sideslipping landing approach in a crosswind. The ailerons bank the plane into the wind (to the right) to stop lateral drift, and the rudder (deflected to the left) keeps the fuselage aligned with the runway. *Inset:* In a crabbed approach, the wings are level, but the fuselage angles into the wind.

judgment at a crucial instant. If the pilot misjudges that final swing, he may touch down misaligned with the runway and risk running off the pavement or with some sideways drift, which can damage the landing gear.

The other crosswind landing technique is the sideslip approach. Here, the pilot uses the ailerons to bank into the wind just enough to counter the sideways drift and employs an opposite rudder to keep the nose aligned with the runway and prevent the plane from actually turning into the wind (fig. 5.4). Although the pilot might need to use a higher throttle setting or a faster-than-normal approach speed to off-set the higher descent rate (since the airplane is sideslipping), the airplane is aligned with the runway throughout the approach. Pilots argue about the merits of these two techniques, but most instructors seem to consider the sideslip approach somewhat easier and thus safer for student pilots.[10]

Rudders: Not for the Birds?

Birds and bats have no vertical tails, so they obviously do not have a rudder for yaw control. How do they deal with adverse yaw in banked turns? As best as I can tell, this is currently an open question, but I suspect they use at least a couple different tricks. British researchers Adrian Thomas and Graham Taylor suggested that birds can improve their yaw stability by twisting their tails, so presumably they can also use twisting to adjust yaw.[11] For example, movies of large gliding birds in gusty air often show them making rapid, twisting tail motions. Most birds, however, can fly even with their tail feathers clipped off, so they must not depend too heavily on tail twisting.

Birds may also use a motion similar to what airplane designers call "aileron differential." Some airplanes have ailerons that swing up much more than they swing down: a lowered aileron produces more drag than does one that has been raised by the same amount. So if the ailerons can be adjusted to keep the raised one more deflected than the lowered one, adverse yaw can be reduced.[12] Gliding birds might do the same thing by using differential wing twisting. If a bird twists up the trailing edge of its inner wing more than it twists down the trailing edge of the outer wing, it might be able to equalize the drag on both wings and avoid adverse yaw.

Birds have a few other options as well. For example, simply by turning its head (looking into the turn), a bird can use its head as a rudder to counteract adverse yaw. If it has big enough feet or if they are webbed, it can do something similar with its feet. The flexibility of a bird's body and its parts give it many more options than those available to airplane pilots.

Adverse yaw may not even occur when birds (or bats or insects) turn while flapping. If a flapping flyer produces a bank by flapping harder—with a larger stroke—on one side than the other, the side that produces more lift will produce more thrust as well. Birds could conceivably adjust the size of their stroke and their angle of attack on each side so that adverse yaw is exactly cancelled, producing a coordinated turn.[13] Much research needs to be done before we can confidently say that birds actually do this. But we do know that some flyers have stabilizing reflexes that are no more complex than the reflexes that they would need to coordinate a turn. Birds, bats, and insects obviously turn just fine, thank you, so evolution has clearly equipped them to produce reasonably well-coordinated turns.

Bird Tails Are More Than Elevators

British biologist Adrian Thomas has carefully analyzed the aerodynamics of bird tails and used his analysis to predict how birds should use their tails. The flight behavior of birds fits his predictions nicely, so his theory seems to be an excellent explanation of how bird tails work. Thomas has found that bird tails operate quite differently from airplane tails. According to him, they act mainly to adjust the bird's overall lift and drag and much less to control maneuvering. A widely spread tail increases both lift and the lift-to-drag ratio compared to the wing alone. Moreover, by adjusting the tail's angle (trailing edge up or trailing edge down), a bird can precisely adjust its total lift, which can be handy during sharp turns or other maneuvers.[14] In fact, a widely spread tail may help delay stall of the wing to unusually high angles of attack so that the bird can fly even more slowly.

At high speeds, the drag of a widely spread tail becomes inefficient; as a result, the bird collapses the tail feathers together so that they produce little lift or drag. This works because, at high speeds, the wing alone produces plenty of lift and the bird doesn't need the extra lift from its tail.

In some maneuvers, bird tails take on the functions of both elevators and wing flaps. When a large bird such as a vulture approaches a landing spot, she first tilts up her widespread tail, which pushes down on her back end and increases her wings' angle of attack, in the same way that an elevator functions on an airplane. As the bird slows down, she tilts down the tail like an airplane wing flap to increase her total lift. Finally, just before touchdown, she bends her tail almost straight down so that the tail produces lots of drag, similar to an air brake, to kill her remaining forward speed.[15]

Although birds with plucked or clipped tail feathers have trouble taking off and landing, they seem to fly more or less normally at cruising airspeeds. As a result, biologists for many years assumed that birds do not use their tails for maneuvers such as turns. From his aerodynamic analysis, Thomas reached a different conclusion. He showed that by twisting a triangular tail right or left, a bird is able to produce both yawing and rolling forces. Such a twisted tail can actually combine some of the properties of ailerons and rudders simultaneously. Thomas suggested that using the tail this way is probably most important for fine-trim adjustments at low speeds rather than for primary direction control.[16] Still, by twisting their tails, birds have apparently discovered a way to get at least some of the effect of a rudder without having to put up with the constant drag cost of a vertical tail surface.

When I first read Thomas's description of this effect, I was not at all surprised. Anyone who has watched close-up videos of a soaring hawk or a vulture on a television nature show can see that soaring birds obviously twist their tails a great deal in slow flight. In fact, they typically twist their tails both in turns and during reaction to gusts. These large birds certainly use their tails to help them maneuver at low speeds, as Thomas predicted, but whether small birds do the same remains to be seen.

An airplane's tail typically provides most of the craft's passive stability (see Chapter 4), but bird tails play a different role in stabilizing bird flight. Given that birds don't seem to need their tails in cruising flight, the tail must not have a role in stabilizing fast, forward flight. Instead, the behavior of soaring birds suggests that their tails operate as part of their active stabilizing reflex system. If a gust unexpectedly tips the bird to one side, it twists its tail to roll back upright. The tail alone may be sufficient to recover from small disruptions, while the bird may need to use aileron-like wing twisting and tail twisting to recover from larger perturbations. Because birds tend to collapse, or furl, their tails at higher flight speeds to reduce drag, the tail probably only plays this stabilizing role at low speeds.

Doing without Tails

A number of airplanes have been built without tails, some achieving more success than others. In contrast to most birds, bats are aerodynamically tailless as well. How do airplanes, which usually depend heavily on tails for stability, get by without them? Do bats, or even birds, have anything in common with such airplanes?

As early as 1799, British scientist and inventor Sir George Cayley drew up designs for flying machines with what we now recognize as conventional airplane tails; so airplane designers who envision airplanes without tails are going against a very long-established tradition.[17] Why give up the natural stabilizing effect, not to mention the control methods, of a conventional tail? The holy grail of tailless airplane design is inherently lower drag, which can lead to either higher speed or longer range (or both).[18]

Tails produce drag in two ways. First, just as any surface that sticks out from an airplane experiences drag, an airplane's tail surfaces add to the total drag on the craft. Second, tails produce trim drag. Trimming means making an adjustment that effectively changes the neutral position of the elevator (and sometimes the rudder). Pilots typically must trim airplanes for different flight regimes: climb, slow cruise, fast cruise, descent, and so on. Usually the elevator is slightly deflected at these various trim settings, and any such deflection further increases the drag of the tail. If a designer can do away with the tail, the resulting airplane should in principle have substantially lower overall drag.

The problem, of course, is that an airplane with a conventional wing but without a tail can be unstable in pitch and yaw. In addition to losing the weathervane effect of a tail, such airplanes have additional pitch instability from an unfortunate property of conventional airfoils. The center of lift on most wings moves around as the wing's angle of attack changes. Just as an airplane has a center of gravity (an imaginary point where the craft behaves physically as if all its mass were concentrated), a wing has a center of lift—a sort of average lift location, a point that we can treat as if the lift were concentrated there rather than being distributed over the whole wing. If the center of gravity and the center of lift are very close together, the wing tends to maintain its orientation. On cambered wings, however, if the angle of attack increases, the center of lift moves forward; and if the angle of attack decreases, the center of lift moves back. Imagine a wing, all by itself, flying along nice and level with its center of gravity and center of lift neatly overlapping. Now the wing increases its angle of attack. The center of lift moves forward, ahead of the center of gravity, giving the wing a further tendency to tilt nose-up, which further increases the angle of attack. The center of lift moves even farther forward, pulling the front of the wing up even harder, increasing the angle of attack even further. This vicious cycle continues until the wing stalls.[19]

A horizontal tail behind the wing efficiently prevents this runaway increase in angle of attack because, as the main wing's angle of attack increases, so does the tail's angle of attack. At a higher angle of attack, the tail produces more lift, pulling the tail up and forcing the nose down. This upward pull of the tail automatically compensates for the nose-up pull of the main wing.

Without a tail, an airplane needs to use various modifications to its wing to regain some stability. One possibility is to modify the airfoil, giving it a slight upward bend just before the trailing edge. This is called a reflexed airfoil, or giving the wing reflex (fig. 5.5).[20] Another is to use swept wings, especially combined with a twist called washout.[21]

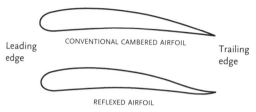

Leading edge   CONVENTIONAL CAMBERED AIRFOIL   Trailing edge

REFLEXED AIRFOIL

Figure 5.5 Conventional cambered airfoil (top) and airfoil with reflexed camber (bottom). Note that the trailing edge of the reflexed airfoil bends up.

Swept wings (also known as wings with sweepback) are angled back from the root to the tip. Washout is a lengthwise twist, nose down at the tips, that gives a wing slightly lower angles of attack at the tips and higher toward the center. A third technique is to droop the tips, or bend down a short section of the outermost wing.[22]

Ironically, once biologists knew what to look for, they discovered that Mother Nature had figured out these tricks millions of years earlier. This is a typical pattern of discovery in biomechanics. Rather than learning tricks from nature, engineers often first figure out some beneficial process or configuration. Armed with knowledge of this benefit, such as the optimal speeds that soarers should fly between thermals (see Chapter 7) or the stabilizing effects of various wing shapes, biologists then look for animals that are taking advantage of such benefits. As often as not, biologists discover that natural selection had produced the same or similar benefits millions of years before engineers had figured them out.

Tailless in Nature

Both bats and birds use some of the tailless tricks. Bats are effectively tailless: most have a short, rodent-like tail embedded in a membrane stretched between their hind legs. This membrane, or uropatagium, functions somewhere between a short bird's tail and a simple extension of the trailing edge between right and left wings. Gliding bats use both reflexed wings and drooped wing tips.[23] Some birds, especially falcons, sweep back their wings and add washout when they make fast glides, a situation when they usually also have the tail furled to reduce drag.[24]

Igo Etrich, an experimenter from the Austro-Hungarian Empire (now the Czech Republic), was involved in an extremely rare case of

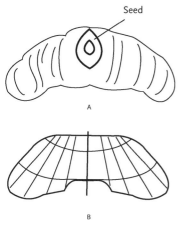

Seed

A

B

Figure 5.6 Top views of (A) an *Alsomitra* ("*Zanonia*") gliding fruit and (B) Igo Etrich's tailless glider. (The pilot was suspended under the wing of Etrich's craft.)

successfully borrowing a design directly from nature. Etrich had learned about the amazing gliding ability of the seeds of a tree in the rainforests of Southeast Asia and New Guinea. Now known to scientists as *Alsomitra macrocarpa*, the tree was known in Etrich's day as *Zanonia macrocarpa*, a name still used by engineers and historians.[25] *Alsomitra* does not have an English common name, although botanists dubbed its samara (the winged seed, or fruit) the "Javanese flying cucumber." The "flying" part makes sense, but the samara has little in common with a cucumber. The seed itself is about the size of a large coin and is surrounded by a stiff membrane that forms a wing with a span of about six inches and a chord of about two inches. This "craft" has no tail and is perhaps the best example of a pure flying wing in nature. *Alsomitra* samaras are very stable gliders, which they must be because they have no nerves or muscles and thus lack any sort of active control mechanism. *Alsomitra*'s wing has both reflex and sweepback, both of which contribute to its remarkable stability.[26]

Etrich decided to build a glider that looked exactly like a scaled-up *Alsomitra* samara: it had the same general outline, with swept wings and reflex at the rear and no hint of a tail (fig. 5.6).[27] He started building his glider in 1904 and was making successful glides with it by 1906. For straight-line glides, the *Alsomitra* design was great, requiring little pilot effort or skill. Etrich then added an engine and discovered the drawbacks of trying to steer a tailless design. Abandoning the tailless approach, he went on to build a series of graceful, tailed airplanes. He continued to be inspired by nature, and his airplanes were clearly strongly inspired by birds. Etrich called these airplanes *taube* (German for "dove"), and they were reputedly easy to fly by the standards of the day.

Tailless Technology

Tailless airplanes largely fit into one of two categories: first, airplanes with wings mounted on a short, stubby fuselage; or second, pure flying wings, with cabins, engines, and so on, mostly contained in a thick center-wing section. (A possible third category, delta-winged craft such as the Concorde supersonic airliner, are really semi-tailless: they usually have a normal vertical tail and flight characteristics closer to conventional airplanes than to flying wings.) The Westland Pterodactyls of the 1930s belong in the wing-and-fuselage group. The Pterodactyls were prototypes designed to investigate whether or not a fighter based on such a design would have lower drag or better visibility than conventional designs. Most versions were really semi-tailless because they had vertical surfaces with rudders on each wingtip, although at least one version was completely tailless. Limited performance and resistance to their unconventional design kept Pterodactyls from going into production.[28]

Before and during World War II, Germans Walter and Reimar Horton built a series of pure flying-wing airplanes, most of which they flew as gliders. They built at least two propeller-driven fighter prototypes. Their last flying wing was intended to fly with jet engines as a fighter prototype. Although their jet-powered craft flew briefly, wartime conditions in Germany prevented them from seriously testing this last machine or putting it into production.[29]

In the United States, the Northrop Company built several pure flying-wing research aircraft during and after World War II. Company founder, John K. "Jack" Northrop, was convinced that the drag advantages and simpler, lighter structure made pure flying wings the wave of the future. But Northrop's propeller-driven XB-35 and jet-powered YB-49 heavy bomber prototypes never quite overcame their stability problems so did not enter production.[30]

Modern, computerized, fly-by-wire control systems have solved such problems. The Northrop-Grumman B-2 Spirit stealth bomber is a pure flying wing that went into production in the early 1990s and currently serves in the U.S. Air Force. Ironically, designers chose the flying-wing configuration mainly for its radar-evading properties rather than its structural or aerodynamic properties, although this low-drag configuration does give it enormous range (6,000 miles without refueling).[31] Moreover, Northrop-Grumman, the B-2's prime contractor, was formed from a merger involving Jack Northrop's company. So far, however, the promise of better aerodynamic efficiency and structural simplicity has not led to a successful civilian tailless design, although many flying-wing designs for airliners have been proposed over the years.

~ 6 ~

Flight Instruments

One fall morning on a rural Wisconsin freeway near Milwaukee, unexpectedly thick fog triggered a pile-up that eventually involved thirty-eight vehicles. Needless to say, none of them were airplanes. Yet even though airplanes don't usually get caught in such chain-reaction accidents, poor visibility nonetheless takes a toll. For example, in the 1990s, dozens of Alaskan bush pilots inadvertently flew their small planes into poor weather and subsequently crashed. Nor are animals immune: every year, thousands of migrating birds are killed when they fly into radio towers or wind turbines at night or in poor visibility.

Whether I am driving or flying, I need some way to know where I am going. Locomotion is not very useful unless I can tell where I am and where I am headed. In spite of a number of clever devices or tricks that pilots and birds use to find their way, the simplest and most versatile method is to simply look. Our first flight instrument is the venerable, reliable eyeball.

Looking Good

The physics of light makes vision the most precise and effective way to sense where we are and where we are going. As visually oriented primates, we find this very obvious and natural; but in fact, most other mammals rely much more on smell and much less on vision. For a flyer, however, vision is the flight instrument of choice. (Bats are an exception that we will look at separately, but even bats rely on vision whenever enough light is present.) Imagine a bird trying to fly through a forest by smelling or listening for the trees. Sounds painful, doesn't it? If the bird can see the trees, it can detect them much more quickly and with much greater directional accuracy. As a rule, stronger and more agile flyers tend to have the best eyesight.

Airplane pilots consider their eyes to be their best flight instruments as well. When a pilot flies under "Visual Flight Rules" (VFR) conditions, he or she faces the fewest restrictions and is not required to maintain constant contact with air traffic controllers. The conditions that allow for VFR flight vary depending on altitude and proximity to airports or other controlled airspace but typically require visibility of at least one mile and the ability to stay more than 1,000 feet away from clouds. The pilot alone is responsible for watching out for other traffic, and the primary rule is "see and be seen." If the weather does not meet these minimum criteria, then pilots are in meteorological conditions that require them to fly under "Instrument Flight Rules" (IFR). In this case, they face many more rules and restrictions and generally must be in direct contact with air traffic controllers.[1]

Pilots use vision for basic flying ("Am I climbing or diving?" "Am I level or turning?") as well as for navigation: "Follow the interstate until I see the lake. Turn right and fly until I see the water tower. The runway will be off to the left." Navigating by flying from one landmark to the next (called *pilotage*) is the oldest form of aerial navigation. Certainly the vast majority of flying animals navigate by pilotage, and they probably have since animals first acquired flight many millions of years ago.

Some flying animals, particularly aerial predators such as falcons and dragonflies, have sharp visual acuity and generally well developed eyesight like our own: binocular vision (for measuring distances), good near and distant vision, color vision, and high sensitivity to motion.[2] The cost of such good vision is that an animal's brain requires a great deal of processing power just for sight. For example, the visual processing areas of a dragonfly's brain take up more than 80 percent of the brain's total volume.[3] Insects tend to be more extreme in this area than vertebrates are, although humans devote at least one-third of our brains to vision, including more than half of the cerebral cortex (the main processing region) for visually complex tasks.[4] Clearly, sharp vision is handy for aerial predators, whether the animal is a dragonfly snatching a mosquito in mid-buzz, a falcon diving at a starling at more than one hundred miles per hour, or a red-tailed hawk soaring 1,000 feet above a meadow looking for chipmunks.

Of course, vision also helps prey species to spot predators in time to avoid being eaten. Prey species often benefit from different aspects of vision, however. Rather than sharp binocular vision along with an ability to see fine detail, they benefit most from wide-angle vision because detecting moving objects (potential predators) is more important to them than seeing fine details. Mallard ducks illustrate this pattern nicely. With eyes facing more to the side than to the front, these birds have severely limited binocular vision; but they can see in all directions, even behind their heads, without moving their heads or eyes, and their eyes and brains are extremely sensitive to anything that moves in their visual field.

Insects, with their compound eyes, often have different specializations in different parts of the eye. Each eye is made up of hundreds or thousands of immovable eyelets, or facets. Each facet looks in a slightly different direction from its neighbors, and individual facets don't form a complete image. Rather, each sends a general visual impression to the brain. In the brain, an image is formed by assembling the signals from each facet as one might assemble the tiles of a mosaic. Different facets can have different optical or functional properties in different places on the same eye. Some bees, for example, have facets that look downward and monitor the insect's progress over the ground as they fly. Bees use those downward-looking facets to measure their speed and how far they have flown. Bees and flies also have front-facing and side-looking facets that help steering.

If a fly's facets on the right side report that objects are moving by quickly, while the facets on the left report that objects are moving by more slowly, the fly senses a turn to the left and adjusts its wings to correct its course back to the right.[5] In many insects, however, facets that look forward or upward do not trigger these course corrections. Instead, the forward-looking facets often provide binocular vision and sharper image formation. Nevertheless, because of their mosaic nature, the sharpest insect eyes are much less keen than a hawk's or a human's eye.[6]

Honeybees need to be able to find patches of flowers at distances of more than a kilometer from the hive and to return that same distance carrying loads of nectar and pollen. They do this by using their down-ward-looking facets to measure how far they have flown.[7] One curious aspect of this mechanism is that honeybees' ability to measure distances can be fooled. If the bee flies lower than normal, the ground appears to move faster; but if the bee flies higher, the ground appears to move more slowly. You will recognize this phenomenon if you look out the window of an airliner on a clear day. The ground appears to move very slowly, even though you are flying more than eight times faster than a fast automobile moves. Your distance from the ground creates the illusion that the airliner is moving slowly compared to your sense of speed in an auto. The bee's flight altitude has the same effect. Bees' distance-measuring system is apparently calibrated for some intermediate height, and strangely, they don't seem to be able to adjust it if they fly very high or very low. When biologists force bees to fly a few inches above the ground, the bees behave as if they have gone much farther than they really have.[8]

As good as it is, vision obviously does not work all the time. At night or in heavy rain, fog, or dust storms, eyes don't work. If I depend on vision to fly, I am grounded when I cannot see. Indeed, more than a few flyers, both human and animal, are stuck on the ground when they cannot see. A fair number, however, do manage to fly in the dark or in low visibility; and some pilots even have devices that allow them to "see" much farther than their eyes would permit. How do these flyers get around when their eyes are all but useless?

Blind As a Bat

Bats are not really blind, but most of them fly at night, and many live in totally lightless caves. They seem to fly just fine in the dark. In fact, they "see" with their ears in the dark. Bats send out brief, intense pulses of ultrasonic sound—that is, sound at a frequency too high for humans to hear. When sound waves from one of these pulses hit an object, they reflect back to the bats as an echo. The bat hears the echo and can accurately sense the direction and distance to the object. This system is called *echolocation* or sometimes *sonar*, after a device that ships use to detect objects under water. (Sonar is actually an acronym for "*s*ound *n*avigation *a*nd *r*anging.") Blindfolded bats can fly through an array of thin rods or thick wires with no collisions, just by using echolocation. Moreover, they can use echolocation to detect flying insects with enough precision to attack and catch them in flight.[9]

Why do they use such high-frequency sounds, which are very strenuous to produce? Compared to medium-frequency echoes, high-frequency sounds can be better aimed and can detect smaller objects. As frequency goes up, wavelength goes down, and the wavelength must be smaller than the dimensions of the object for the object to produce a good echo. Ultrasonic frequencies can give echoes from objects as small as insects.[10] Bat echolocation has a shorter range than good vision and is not quite as sharp, but it is a close second. It is probably similar to peering through a thick fog: nearby objects are reasonably easy to sense, but objects more than a few dozen feet away fade into obscurity. For an animal flying in the dark, where vision is not much use, echolocation is by far the best detection system that animals have evolved. And despite our common use of the phrase "blind as a bat," most bats have reasonably good vision. On moonlit nights or at dusk and dawn, they use their eyes and navigate visually, although they may still use echolocation to find and attack insect prey.

Many airplanes use radar (an acronym for "*r*adio *d*etection *a*nd *r*anging"), an echo-ranging system that works in ways similar to bat echolocation. At about the same time that researchers were developing sonar to detect submarines underwater (before World War II), others were developing a radio-based system to detect airplanes flying at great distances.[11] Radar systems project a tightly focused beam of microwave radio energy that sweeps through the environment and then "listen" to detect any radio waves that reflect back from solid objects. They work just as bats do, identifying the direction and distance of an object by the direction and timing of the echo. The big advantage of radar over vision is that radar works across much greater distances, typically tens to hundreds of miles, depending on the power and sophistication of the system. Moreover, radar works in the dark and can even "see" though some clouds. It is in many ways a major improvement over a pilot's vision.[12]

Radar has several uses in aviation. Fighter-interceptor airplanes use it to locate, track, and target airplanes at great distances. Another military use is to locate objects on the ground, either for navigation or to identify and aim at targets. On the civilian side, air traffic controllers use ground-based radar to keep track of aerial traffic over huge volumes of sky, hundreds or thousands of times more than they could handle with binoculars. Weather radar is by far the most common type of radar carried on civilian airplanes. Weather radar units can detect raindrops, hail, and, to some extent, even turbulence in storm clouds. Especially at night, they give a pilot advance knowledge of whether or not a given cloud or storm might be too dangerous to fly through.[13]

As a rule, radar designed for one purpose does not work well for a different purpose: weather radar cannot be used to find objects on the ground, for example, nor can radar for imaging the ground be used to detect other airplanes in flight. There are more sophisticated radar systems that can do more than one task, such as locating both ground and aerial targets. They do so mostly by using different kinds of microwave beams (different frequencies or intensities or even different antennas) for different tasks. Bats have a big advantage in this regard. Yes, they use pulses with different characteristics when searching for aerial prey as opposed to zeroing in on a target insect, but they broadcast both kinds of pulses with their voice and detect and analyze both kinds of echoes with their ears and brain. The versatility of their senses and their nervous system gives them much greater flexibility than any electronic device can provide. They have the advantage of being able to do several different detection tasks with one set of "equipment."

A Compass to Point the Way

Aside from the new and versatile global positioning system (GPS) receiver, the magnetic compass is the device we associate most closely with navigation. Helmsmen have been steering ships using magnetic compasses for many centuries, and aviation adopted compasses as soon as pilots began flying farther than they could see. A magnetic compass detects the magnetic field of the earth, and a compass needle always points toward the magnetic North Pole. Using a pocket compass is easy: line up the needle with the "north" or "0 degree" marker on the compass face; find the degree marking for the direction in which you wish to go; and keeping the needle on the "north" marker, proceed in the direction of the degree marker you have chosen. (Because the magnetic pole is not located at the true North Pole, a correction called *magnetic variation* must be added to or subtracted from the compass reading to find true north in most locations; the size of this correction is shown on high-quality topographic maps and aviation charts.)

Lubber Lines and Liquor

Marine and aviation compasses are different from the typical pocket compasses used by hikers and more intuitive to use. The magnet is attached to a disk or ring that shows compass bearings; and as the compass case rotates, a mark or line on the case (called the *lubber line*) aligns with the compass bearing as shown by the disk or ring (fig. 6.1). The helmsman or pilot simply reads the compass bearing that aligns with the lubber line. The compass case is attached to the craft, so the compass bearing under the lubber line gives the craft's heading.

Of course, navigation is more complicated than just following a particular compass bearing. A compass is useful only if you know which heading to take and for how long, so navigators generally use

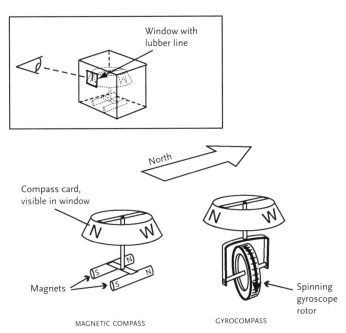

Figure 6.1 The working parts inside of two types of compass. The magnetic compass uses magnets to stay aligned with the Earth's magnetic field. The gyrocompass maintains its orientation due to the gyroscopic effect of its spinning rotor and must be manually aligned with north. *Inset:* How the user reads these compasses.

both compasses and maps. Marine charts and terrestrial maps have been around for centuries, but the aeronautical chart is a relatively recent development. Aeronautical charts combine a topographic map with symbols to show the elevation of potential obstacles (such as radio towers, skyscrapers, and mountains) and the location of navigational aids and restricted airspace.

Aviators discovered early on that magnetic compasses are subject to a peculiar set of errors unique to aircraft. In the northern hemisphere, the lines of the Earth's magnetic field point downward as well as north. This tilt, combined with the way in which an airplane banks to one side in a turn, causes the compass to give incorrect readings during a turn to the north or the south. That same tilted magnetic field, combined with the compass needle's inertia, causes incorrect readings when the airplane accelerates or decelerates on east or west headings.[14] Needless to say, a pilot needs either lots of practice and skill or several tries to roll out of a turn exactly on a given magnetic compass heading. In fact, this confusing compass behavior has led to widespread use of the gyrocompass, or directional gyro, in conjunction with magnetic compasses, even on small, simple airplanes. We'll look more at gyroscopic instruments shortly.

To differentiate between gyroscopic and magnetic compasses, pilots often use the terms *DG* for directional gyros and *whiskey compass* for magnetic compasses. Why "whiskey"? Mariners learned long ago that, by filling the case with liquid, they could keep the compass card or needle from bouncing around erratically with every bump or vibration. Water is no good because, if the air temperature drops below freezing, water will freeze and burst the case. Moreover, the iron magnet and other iron parts tend to rust in water. Because alcohol is a clear liquid and does not freeze at low temperatures, it was traditionally used in marine compasses—hence, the "whiskey" description. Early aviators adopted the same practice, but you would not want to drink the liquid out of a modern aviation compass. Nowadays, kerosene or light oil is used. These liquids are more viscous than water or alcohol, so they dampen erratic movements better. Moreover, they lubricate the pivot and they do not cause the magnet or other steel parts to rust.

Animal Magnetism

Biologists have known for several decades that birds can use the sun and stars for navigation, but they were startled a couple of decades ago to discover that many migratory birds seem to have an internal magnetic compass. Even if they cannot see the sun or stars, these birds can fly in a desired direction. Moreover, when scientists put these birds into magnetic fields that were artificially rotated by ninety degrees from the Earth's normal magnetic field, the birds tried to fly at right angles to their normal migratory direction.[15] We now know that a number of other animals, including lobsters and honeybees, also have internal magnetic compasses that they use for navigation. Biologists have been unable to locate specific sense organs that detect magnetic fields, although some brain cells seem to contain magnetic particles; so the compass may reside in the brain itself. At this point, the evidence is still too weak to explain where and how animals sense the Earth's magnetic field.

Birds use their magnetic compasses very differently from the way in which we use ours. We hold our compasses flat so that the needle swings horizontally. But the magnetic force lines of the Earth's magnetic field are not horizontal. Recall the problem this causes for banked turns and accelerations in airplanes. Birds act as if their magnetic compass needles swing vertically instead of horizontally, in the same way that a surveyor's vertical dip compass works: the magnetic field lines point downward toward the magnetic North Pole in the northern hemisphere, upward away from the magnetic South Pole in the southern hemisphere, and horizontally at the equator. In other words, a dip compass will tell you whether you are going toward or away from the pole, except at the equator, when it tells you that you are at the equator but gives no directional information. This is exactly how birds respond. First, they cannot orient with their magnetic sense in magnetic fields that are exactly horizontal, as at the equator. Second, in experimentally manipulated magnetic fields that point up and to the north (instead of down and north as in the Earth's normal field), birds act as if the directions of the poles and the equator have been reversed, whereas a whiskey compass would still point north in such a modified field.[16] Most migratory birds migrate along north-south routes, and their internal compass tells them when they are flying toward or away from one of the poles.

Because the angle of the field gets steeper the closer you get to one of the magnetic poles, a dip compass is helpful because it can tell you how close you are to one of the poles or the equator. On the other hand, even under the best conditions, its heading information is not as precise as that of a conventional horizontal compass; and at the equator it is useless for directions. But for birds, the advantage of latitudinal information ("How close am I to the equator?") more than compensates for the lack of directional accuracy.

One way to use a compass is to combine it with a map and landmarks, as hikers and backpackers do when they get away from roads.

Another way is to move a known distance on a known compass heading to end up at a particular spot, regardless of landmarks, as in the
sport of orienteering: for instance, "Proceed 250 yards on a compass
heading of 37 degrees. Turn to a heading of 186 degrees and move 120
yards. Then turn to a heading of 101 degrees and move 310 yards." In
a ship or an airplane, the known distance is replaced by a known time
at a known speed, a method called *dead reckoning*. Dead reckoning is
certainly useful for navigating when you cannot see the ground, but it
won't tell you if you are climbing or diving or if you are flying right
side up or upside down. Flying in low-visibility conditions requires
more sophisticated instruments.

Flying Blind: How to Fly If You Can't See

All flyers fly at the mercy of the weather. In bad storms flying animals
and prudent pilots don't fly. But what about in the dark or in fog or
non-storm clouds? Once airplanes became reasonably reliable, aviators wanted to be able to fly in low-visibility conditions. But those who
tried usually became disoriented or succumbed to vertigo and crashed.

Using our eyes and inner ears, we humans can tell which way is up
and how we are moving. If we can't see, we depend almost entirely on
the motion and gravity sensors in our inner ears (the vestibular system), but these sensors are not adapted to flight. In fact, our inner
ears' vestibular systems are easily tricked. Sitting in an airplane in a
steady, coordinated turn, our inner ears confidently report that we are
not turning at all. (Try it yourself: next time you are on an airliner,
close your eyes during a turn and see if you can feel any sense of turning.) Upon straightening out from a turn, our inner ears tell us that
we are actually turning the other way rather than stopping the turn.
Clearly, we cannot depend on our own senses when we fly in low-
visibility conditions; and those who try almost invariably crash.

Researchers in the 1920s and 1930s put great efforts into developing flight instruments that would allow pilots to fly safely in the dark
or in clouds. These instruments mostly rely on gyroscopes. In 1929,
James H. "Jimmie" Doolittle (better known for his daring bombing
raid on Tokyo early in World War II) made the first airplane flight

entirely on instruments, from take-off to landing. He flew under a hood that completely blocked his view out of the airplane and used a variety of newly developed instruments, including a gyroscopic artificial horizon. A safety pilot flew along in a normal cockpit as a precaution but did not touch the controls.[17] Since the 1940s, airplane flights made entirely on instruments have become commonplace, although in the United States, IFR requires the pilot to see the runway before he or she is allowed to descend below some minimum altitude—typically three hundred to five hundred feet above the ground, depending on the airport. Oddly enough, the auto-land function of the autopilot on most modern airliners is perfectly capable of landing the airplane safely with no visibility at all, and the pilot also has instruments that are sensitive enough to carry out a manual landing in the same conditions.[18] So far, however, the Federal Aviation Administration has been reluctant to give any airports approval to allow the near-zero visibility landings it calls "Category IIIc," largely because of the problems of detecting and avoiding other airplanes already on the ground.

A gyroscope is basically a disk with a weighted rim that spins very rapidly. When such a disk is spinning, it resists changes in the orientation of its axle. In other words, if you try to tilt it, it pushes back, although it pushes back with a torque or turning force at right angles to the original push.[19] Toy gyroscopes that perch at seemingly impossible angles are using this property: if they spin fast enough, their resistance to falling over is greater than their weight. With gyroscopic instruments, the idea is that either the gyroscope is in a set of pivoting rings called gimbals that allow it to stay fixed in space as the airplane turns around it or the gyroscope is rigidly connected to the airplane and the instrument measures how hard the gyroscope pushes on its mounting when the airplane turns (fig. 6.2).

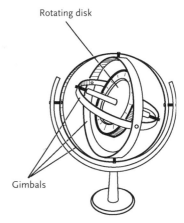

Figure 6.2 A gyroscope (spinning disk) in a set of gimbals, allowing the gyroscope axis to maintain a constant orientation even if the supporting base changes its orientation.

The gyrocompass, or directional gyro (DG), is one of the simplest gyroscopic instruments. The pilot adjusts the instrument so the axis of the gyroscope disk points north; then the gyroscope swings freely in a gimbal (fig. 6.1). As the airplane turns, the gyroscope continues to point north, and the rotation of the case around the gyroscope drives an indicator that shows the airplane's heading. The DG does not suffer from the errors of turning or accelerating that plague magnetic compasses, making it much more reliable when planes are maneuvering.[20]

Why not dispense with magnetic compasses altogether and rely on DGs? One reason is that the magnetic compass is entirely passive, whereas the DG requires some power source—electricity or engine vacuum—to keep the disk spinning. The whiskey compass is the ultimate emergency backup. More important, due to the Earth's rotation and friction in their bearings, gyroscopes tend to drift very slowly. This is called *precession,* and it means that the pilot must periodically (ideally, about every fifteen minutes) reset the DG using the magnetic compass. Of course, the airplane must be in straight, level flight at a constant speed, or the reading on the magnetic compass may be wrong.

The turn-and-slip indicator—or its modernized version, the turn coordinator—includes the other type of gyroscopic instrument, a fixed gyro whose axis turns as the airplane turns. The turn-and-slip indicator is really two instruments in one: a gyroscopic instrument that shows how fast the airplane is turning and a simple ball in a curved, oil-filled tube that helps the pilot coordinate ailerons and the rudder in turns. The gyroscopic part of the instrument, which has its axis lined up lengthwise and fixed to the airplane, pushes back against a spring when it is forced to turn as the airplane turns. The faster the airplane turns, the harder the gyro pushes back. The spring is connected to a needle that shows how fast and in what direction the airplane is turning. This instrument takes some training to use effectively because it only shows the rate at which the plane's nose swings to one side or the other, not the rate or angle of bank. A sharp turn at high speed can give the same reading as a gentler turn at low speed, with no indication of how steeply the airplane is banked. The more modern turn coordinator has the gyroscope axis fixed at an angle so that when the airplane first rolls to one side, the instrument indicates how fast it is rolling. Once the airplane has reached a steady bank angle and stops rolling, the instrument shows the rate of turn, just as the turn-and-slip indicator does. The turn coordinator thus gives more information, but it is still not particularly intuitive to use and requires training and practice. It is, however, a simple, reliable instrument and can be found in even the smallest and simplest of airplanes.[21]

At the other end of the complexity scale is the artificial horizon, or attitude indicator (AI).[22] This instrument contains a gyroscope in gimbals that allow it to tilt freely from side to side and from front to back. The gyroscope is connected to a display that shows a stylized horizon that moves just as the actual horizon appears to move when the airplane maneuvers. If the airplane tilts nose-down, the horizon line moves up. If the airplane banks left, the horizon line tilts to the right. The AI is easy and intuitive to understand, but it is considerably more complex and expensive than either the DG or the turn-and-bank indicator. Only airplanes designed to fly at night or in IFR conditions (or training aircraft) need an AI, so builders of small, simple airplanes often do not bother installing them.

Animals have evolved at least two "flight instruments" that help them fly when they cannot see, one of which rivals the gyroscopic attitude indicator in effectiveness. The first of these uses the insect's head as an indicator. Dragonflies have rather large heads as insects go. A dragonfly's "neck" is basically a loose pivot connecting the upper back of the head to the thorax. The link is similar to the way in which a picture hangs on a wall: the head is free to swing left and right like a pendulum. When a dragonfly flies, gravity tends to hold the head upright, even if a gust tilts its body or it inadvertently starts to bank. When it senses that its head is at a different angle from the body's, the insect adjusts its wing beat to correct for the tilt.

More than fifty years ago, German researcher Horst Mittelstaedt ran some clever experiments to figure out this mechanism. He attached

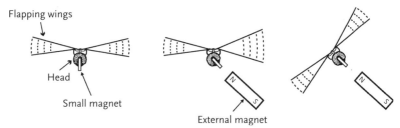

Figure 6.3 Mittelstaedt's experiment showing how a dragonfly's head helps to stabilize its flight (front view). As long as the head is upright, it flies straight (left). When a magnet tilts its head (center), the dragonfly adjusts its wing beat to bring its body back into alignment with its head (right).

dragonflies to a tether that allowed them to roll from side to side and tricked them into flapping by blowing an air current on them. He then glued tiny magnets to their heads and used a large magnet to tilt the heads from a distance without actually touching them. He discovered that dragonflies flapped their wings to roll their bodies to match the tilt of the head (fig. 6.3).[23] This reflex acts as a potent stabilizer for the dragonfly, but it is not foolproof. In a properly coordinated turn, the head is tilted by centrifugal force so that it stays aligned with the body, meaning that a dragonfly cannot use its head to sense the difference between straight flight and a coordinated turn. Presumably, however, coordinated turns are intentional, so this sensory gap does not seem to cause dragonflies any trouble.

Flies have evolved an even more remarkable and sophisticated system. This large and diverse group of insects, which includes house flies, horse flies, crane flies, deer flies, bluebottle flies, midges, and mosquitoes, has evolved what amounts to an artificial horizon. The front pair of wings performs its usual aerodynamic function, but the hind wings have been reduced to tiny knob-tipped stalks called *halteres*.[24] As the wings flap, the halteres also swing up and down, and the knob on the end of each haltere acts as a vibrating mass. When a vibrating mass is rotated, a phenomenon called the Coriolis effect produces a force at right angles to the plane of vibration, much like a gyroscope. (You can feel this yourself with a large tuning fork: strike it to start its vibrations; then tilt it sharply in various directions to feel how it resists those movements.) As a fly's body tilts or rotates, the halteres bend a little. Nerve endings in the halteres sense this bending and trigger a correcting reflex in the flight muscles.[25] With this system, flies can fly in a straight line even in total darkness, although they still need to be able to see to avoid obstacles and find a landing spot.

Biologists first demonstrated the importance of halteres by the simple but crude tactic of cutting them off. Without halteres, a fly can often fly but rarely for very long. Usually, within a second or two, the fly enters what pilots call a death spiral: a descending turn that gets tighter and steeper until the fly hits the ground. (The same thing happens to pilots who try to fly in clouds without gyroscopic instruments or appropriate training.) If the experimenter glues a length of light string to the back end of the fly, the thread acts as a stabilizer, like the tail of a kite. The fly can fly again, although not with its normal agility. Entomology teachers have been using this demonstration for decades, thus bearing out entomologist Vincent Dethier's observation that young boys who don't grow out of pulling the wings off flies either come to a bad end or become biologists.[26]

Humans have developed a striking technological analogue to the fly's halteres. This device is called a solid-state gyroscope, or piezo gyro, and I am most familiar with its use in radio-controlled helicopters. Full-sized helicopters are difficult to fly because of their lack of inherent stability and their ability to fly in any direction, including sideways and backwards (see Chapter 8). Radio-controlled helicopters are even more challenging: because of their small size, they change direction even faster, and the pilot does not have the sensory feedback of being in the cockpit and moving with the helicopter. In the 1970s, a few intrepid souls built R/C helicopter models and, with near heroic efforts (including many expensive crashes), learned to fly them. Within a few years, some people began mounting tiny gyroscopic devices in their model helicopters. These gyros were arranged so that they could sense when the helicopters was yawing (turning nose-left or nose-right). The gyro was connected to the control system almost like an autopilot: when the gyro sensed that the helicopter was yawing, it sent a control signal to counteract the yaw. The pilot could override the gyro when he actually wanted to turn, but the device freed the pilot to worry only about pitching and rolling and made hovering much easier. Because helicopters normally land by hovering, gyros made learning to fly helicopters much easier. (Some experienced R/C pilots eventually dispense with gyros, but almost all start out with them and many continue to use them even after becoming accomplished pilots.)

These early gyros were basically scaled-down turn coordinators; and even though they were miniaturized by airplane standards, by R/C model standards they were heavy, expensive, delicate, and power-hungry. In 1988, a few electronically savvy hobbyists came up with a better idea. They decided to build a gyro based on a vibrating mass rather than a spinning wheel.[27] Piezoelectric crystals change shape when a voltage is applied to them and also produce a voltage when squeezed or bent. An oscillating voltage makes the crystal in a piezo gyro vibrate (like halteres); and if the vibrating crystal rotates in certain directions, it bends at right angles to its vibration and produces its own voltage that indicates how fast it is turning.[28] The device amplifies this signal and uses it to adjust the helicopter controls. These piezo gyros are lighter, cheaper, less fragile, and much less power-hungry than spinning gyros are. So the solid-state gyro in an R/C helicopter uses the same principle as a fly's halteres and handles the same job: adding a stabilizing reflex to an otherwise unstable flyer. Many years later, video camcorder companies developed almost identical devices to stabilize video images, particularly for high-magnification zoom lenses.

How Fast?

If I walk or run, I can tell how fast I am going pretty easily, mainly by how often I take a step. If I drive a car, the speedometer measures how fast the wheels rotate. In both cases, I depend on contact with the ground. Although I use more metabolic energy if I run faster, knowing precisely how fast I run or drive is usually of no great consequence for my locomotion, aside from my desire to obey traffic laws.

Flight is different. By definition it means no contact with the ground, so I cannot "feel" my speed by direct contact with the ground over which I fly. Moreover, knowing my exact speed is much more important when I fly: if I fly too slowly, I stall and fall; if I fly too fast,

I risk structural damage from aerodynamic loads. Of course, as on the ground, if I am trying to go some particular distance, then knowing my speed helps me figure out how far I have gone.

Most airplanes have an airspeed indicator (a gauge or display) using a *pitot-static system*. Consisting of a set of tubes and a pressure sensor, it measures the difference between the pressure of a pitot tube and a static port to determine the airplane's speed. The pitot tube (a short metal tube on the front of one wing or on the nose of the airplane) points forward and senses how hard the air pushes on the front of the airplane as the craft flies—the so-called *ram air pressure*. Pitot tubes are named after their inventor, Henri de Pitot, a Frenchman who studied fluid movements in the eighteenth century (hence, the term is pronounced "PEE-toe"). The static port is a tiny opening on the side of the airplane that detects the local air pressure—that is, the pressure on the airplane at that location as if the craft were not moving, or the static pressure. Designers try to position the static port in a place that is not affected by air currents during flight, which can be challenging. The airplane's airspeed is directly related to the difference between the ram pressure and the static pressure. Airspeed indicators need static ports because the static pressure is not constant; it drops as an airplane goes up in altitude, for example. The operating principle is simple: the instrument pneumatically subtracts the static pressure from the ram pressure and uses either a mechanical or an electronic linkage to convert that pressure difference into an airspeed reading.[29]

The pitot-static system measures airspeed, and the pilot needs to remember that airspeed is not ground speed. Tailwinds increase the ground speed, headwinds slow the ground speed, and crosswinds require heading corrections. When I took flying lessons, I was thrilled that the trainer I flew cruised at a little more than one hundred miles per hour. I will never forget my chagrin at cruising into a headwind above a freeway and barely keeping up with the cars on the ground.

Flying animals need to know their airspeed to avoid stalling and also to help with steering. For birds and bats, this is not much of a problem. They can feel the wind on their faces and wings with the tactile nerve endings in their skin. Bats can even sense tiny patches of turbulence on their wings and make fine adjustments to prevent stalls while maneuvering.[30] Insects, however, effectively wear suits of armor, and their rigid exoskeletons block touch sensing. As a result, insects have no widespread tactile nerve endings and are effectively numb over most of their body surface. To get around this problem, most have bristles or sensory hairs, singly or in patches, scattered over their bodies. These hairs act as touch sensors in particular spots, and some act as wind sensors or airspeed indicators. Grasshoppers, for example, have patches of hair on the front of their heads that act as wind speed sensors.[31]

Other insects use their antennae as wind sensors. Insect antennae are primarily odor sensors (insects have no noses); but because antennae stick out of the front of the head, they are also conveniently placed to sense air currents. Moths, flies, and some beetles use their antennae to sense airspeed and direction as they fly through the air. The faster they fly, the more their antennae bend backward. They "feel" this bending as a measure of airspeed.[32]

A friend once asked me if insects could feel a wind while flying and if it confused them about their actual speed and direction. In fact, an insect (or a bird or an airplane) in flight cannot feel a steady wind. A flyer is fully immersed in the air and is carried along with the movements of the air mass enclosing it. Passengers on hot-air balloons often comment on how still and calm the air is in the basket, even though the balloon is being blown along by a steady wind. Because the balloon is carried along with the moving air, passengers feel no breeze.

The airspeed sensors of airplanes and animals (especially insects) are fundamentally different. A pitot-static airspeed indicator is a pressure-sensing (barometric or pneumatic) device. Insect hairs and antennae, in contrast, depend on deflection or distortion of some body extension. Birds' and bats' wind sensors also depend on deflections, although these are much more subtle than insects' are.[33] This difference is no accident. In principle, I could poke a springy stick out of the side of an airplane and watch how much it bends in the slipstream to measure my airspeed. This device would have at least two drawbacks. First, because of the enormous speed range of most airplanes (multiple hundreds of miles per hour), I doubt if a bendy stick could give me much useful speed information across the whole range of airspeeds. Second, for the stick to work at high speeds, it would not be precise enough at low speeds, and vice versa. The pilot needs accurate information about airspeed when landing so as to avoid stalling prematurely and at high speed to avoid structural damage. A bendy stick can't be accurate at both ends of the range, particularly for large aircraft that operate from one hundred to six hundred miles per hour. The pitot-static system is a mechanically simple, reliable device that can give useful accuracy across a very wide speed range. Flying animals can get by with feeling bristles bend (or with sensitive skin) because they operate over a much narrower range of speeds: less than fifteen miles per hour for insects, perhaps forty to fifty miles per hour for fast birds, less than half that for most birds. Over these speeds, deflection sensors work fine and are much less complex than pitot-static systems.

How High?

One area in which aviation parts company with flying animals is in sensing altitude. Airplanes have one or more sensitive instruments to measure how high they are flying, but nature's flyers don't seem to bother with any mechanism to sense height above the ground. In fact, an airplane pilot often needs to know the airplane's altitude with some precision, but birds and bees generally have little or no use for altitude information. Why the difference?

Human pilots need to know their altitude for at least two reasons: to avoid flying into tall objects or rising terrain and to avoid collisions with other airplanes. Pilots use special maps (aeronautical charts, or sectionals) that show the elevation of terrain above sea level, including hills and mountains, plus the height of tall structures such as radio towers, smokestacks, and water towers. A pilot can plot a route on such a chart and plan to fly at an altitude that carries her at a safe distance above any hills or towers. Naïvely, we might think that such a precaution is unnecessary, especially in good weather during daylight; but in fact, due to high flight speeds and large distances required for turns, dodging or climbing over hills and obstacles is difficult if not impossible. At night or in haze or clouds, flying at a known safe altitude becomes even more crucial.

Pilots also need to know their altitude to avoid other aerial traffic. The Federal Aviation Administration and its predecessors developed a series of "rules of the road" to help pilots avoid collisions. Some were adopted from surface traffic rules, such as "the vehicle approaching from the right has the right of way." Others are unique to flight, including a set of rules for flying at different altitudes depending on the direction of flight. For example, in VFR flight up to 18,000 feet above sea level, airplanes flying headings on the eastern side of the compass fly at one set of altitudes and those heading on the western side of the compass fly at a different set. Airplanes on compass headings of 0 to 179 degrees fly at odd thousand foot altitudes plus five hundred feet—for example, 5,500 or 11,500 feet above sea level. Airplanes heading from 180 to 359 degrees fly at even thousands plus five hundred feet—for instance, at 6,500 or 10,500 feet—thus creating a 1,000-foot differential in altitude between airplanes on opposite headings. (IFR pilots get the even and odd thousand feet altitudes without the "plus five hundred feet" addition.)[34]

An altimeter is any instrument that measures how high an airplane is flying. Standard altimeters are a barometric, or pressure-sensing, device.[35] Other types of non-barometric altimeters fulfill more specialized roles. Standard altimeters work because the air pressure decreases with altitude, as we saw with the airspeed indicator: as the airplane climbs, the pressure drops. In fact, an altimeter is really just a barometer that is calibrated in feet above sea level rather than millibars of air pressure. One complication is that weather patterns also cause changes in the air pressure. The "highs" and "lows" on a weather map refer to high or low air pressure, and weather fronts are usually zones of changing air pressure. So a barometric altimeter has an adjustment knob to compensate for these changes. The pilot checks the barometric pressure before take-off and sets that pressure in a little window on the altimeter to adjust for changes due to weather. The current barometric pressure is one of the standard items of information that an air traffic controller gives to a pilot as the airplane approaches an airport in preparation for landing. This information is especially important because airplanes fly at standard altitudes in the airport traffic pattern, and airplanes would be flying at the wrong altitude if they had incorrect pressure settings on their altimeters.

A barometric altimeter has the annoying property of telling the pilot its altitude above sea level rather than above the ground. Sitting on the ground where I live in Kansas, I am about 850 feet above sea level, and the Denver airport is more than 5,000 feet above sea level. A pilot needs to know how high the ground is along her route so she can stay at a safe altitude above ground level. If I did not understand this and tried to fly at the normal pattern altitude of five hundred feet above ground level by flying at five hundred feet on the altimeter at my local airport, I would hit the ground a couple hundred feet before I got down to five hundred feet on the altimeter.

Many airliners, military airplanes, and other high-performance airplanes have a radar altimeter in addition to the barometric altimeter. The radar altimeter works just like any radar, sending out a radio signal aimed downward; and the beam's reflection off the ground tells the instrument how high the airplane is flying. So the radar altimeter tells the pilot the actual height above the ground rather than the height above sea level.[36] The radar altimeter is no use for traffic avoidance, but it is extremely useful for maneuvering over hilly or mountainous terrain and when landing. If the airplane flies over an unexpectedly high hill or descends inadvertently, the radar altimeter tells the pilot he is getting too close to the ground. In fact, modern airliners have a ground proximity alarm using the radar altimeter that flashes lights and plays audible warnings when the airplane gets too close to the ground. (The alarm is turned off during a normal landing approach, although the altimeter reading itself is still available.)

One final type of altimeter is the sonar altimeter. It works the same way as a radar altimeter but uses an ultrasonic sound pulse (like bats' echolocation) rather than a radio beam. A sonar altimeter is impractical on any typical airplane. Sound does not travel nearly as far or as fast as microwaves do, so a sonar altimeter is not much use at heights of more than about fifty feet. Moreover, airplane engines are so noisy that they would probably swamp the sound pulses of the sonar. The one type of airplane that can make practical use of a sonar altimeter is the human-powered aircraft (HPA). Human-powered airplanes fly low and slow, and their "engine" (the pilot) makes relatively little noise.

An aeronautical engineer named Paul MacCready and his team built an HPA, the Gossamer Albatross, to fly across the English Channel (see Chapter 10). They wanted an instrument that could tell the pilot how high he was flying above the water, which can be very difficult to judge visually from an airplane: are those big waves far away or little waves up close? Because the pilot can barely produce enough power to fly such a craft, keeping it light is even more critical than it is for conventional airplanes. A heavy, bulky radar altimeter was out. Then the team hit on a clever solution. At the time, the Polaroid Corporation was selling a line of auto-focus cameras that used a sonar detector to sense the distance from the camera to the subject in order to set the lens's focus. Polaroid, working with MacCready's team, adapted the detector from a camera by connecting it to a simple digital display. They produced a very cheap, very lightweight, but precise and reliable altitude sensor.[37] This instrument, inspired by bats and airliners and borrowed from a camera, played an important role in the Gossamer Albatross's successful crossing of the English Channel.

As I've mentioned, the altimeter seems to be the one standard airplane instrument that flying animals do entirely without. As far as we know, animals do not have specific sense organs to tell them how high they are flying. Vision (or echolocation, in the case of bats) seems to do the job well enough. Animals rarely fly when they cannot see the ground; and unlike airplanes, they fly slowly enough that they can maneuver to avoid obstacles or hills. Moreover, traffic control is simply not an issue for flying animals. Again, due to relatively low flight speeds and high maneuverability, collisions are rare even in dense flocks. "Rare" does not mean "absent," however. Birds in dense flocks do occasionally collide, particularly during mass take-offs when startled. Because of their low speeds and low inertia (small bodies), these collisions are merely annoying and perhaps painful but not catastrophic. When flocking birds or swarming mayflies accidentally collide, they don't crash; they just fly away.

Our final instrument is a close cousin of the barometric altimeter, and some flying animals may have a use for this one. The rate-of-climb indicator does not sense altitude directly, but it does sense changes in altitude. It measures how fast an airplane goes up or down by sensing how fast the air pressure changes as the altitude changes.[38] In most airplanes, it is a secondary instrument, a backup to the altimeter. In sailplanes, however, where finding rising air currents is crucial (see Chapter 7), these instruments are much more important. Sailplanes use an especially sensitive version called a variometer, which is a primary instrument in the pilots' quest to find lift (rising air) and avoid sink (falling air).

Biologists have not found any specific variometer organs in animals, but animals probably don't really need specific organs to detect pressure changes. If your ears have popped while riding down a mountain road or in a fast elevator, you have detected a pressure change due to a change in altitude. All an animal needs is an air-filled space in the body and normal tactile receptors. Birds and many insects are already well suited to the task because their bodies are full of air sacs that make up part of their respiratory systems. They can certainly feel these air sacs being stretched or squeezed, and I will lay odds that soaring birds such as vultures and eagles pay close attention to how those sacs—and their ears—feel as they fly through rising and falling air currents.

Flying animals can sense many things that we humans cannot. Bats can hear ultrasonic sounds, bees can see ultraviolet light, pigeons can feel the Earth's magnetic field. Biologists have to be clever and insightful to unravel animal sensory systems that the biologists themselves don't possess and so cannot experience directly. We know about several of these properties that animals can sense and that we cannot, but only a foolish person would believe that we have found all of them. More animals using undiscovered, unimagined sensory mechanisms might well be awaiting discovery, and chances are good that some of them will be flyers.

~ 7 ~

Dispensing with Power

Soaring

On any warm, sunny day outside my Kansas town, I am almost
certain to see turkey vultures gliding with their wings rigidly
extended. They wheel in the sky, not flapping at all, rarely descending
yet often climbing. Anyone who visits the U.S. Air Force Academy
near Colorado Springs on a nice day will have a similar experience:
they are almost sure to see the academy's sailplanes soaring along the
front range of the Rocky Mountains. Like the buzzards, they slice
silently though the air, engineless, yet they sometimes stay up for
hours.

Both the turkey vultures and the sailplanes are soaring, or taking
advantage of rising air to stay aloft without power. To understand
soaring, we first need to know something about gliding. Gliding is to
powered flight what coasting downhill on a bicycle is to actively pedaling. Without thrust, a flyer can still fly, but it must fly "downhill"
through the air, letting gravity pull it along rather than an engine or
flapping wings. Indeed, the aeronautical term for a normal descent is
a *glide*, with the term *dive* reserved for excessively steep descents. Any
bird that stops flapping or any airplane with its engine throttled back
can glide. Some machines and animals, however, are specialized for
gliding or even restricted completely to gliding. A surprising array of
animals other than birds, bats, and insects can glide, ranging from

flying squirrels and flying fish to flying lizards and even flying snakes. Likewise, engineless airplanes, called gliders, have been around longer than successful powered airplanes. Otto Lilienthal built and flew gliders more than a decade before the Wright brothers' first powered flight. Moreover, the Wrights taught themselves to fly in gliders before they attempted a powered flight (see Chapter 4).

Soaring is a specialized form of gliding. To soar, a gliding animal or airplane needs to find air rising faster than the glider descends in a glide. If a glider descends at five hundred feet per minute as it flies forward, then it needs to find air rising at five hundred feet per minute or faster. If a glider can find air rising faster than its descent rate, or sinking speed, then the glider can climb higher above the ground, even though it is descending through the air mass it inhabits. This is what buzzards circling over a cornfield or sailplanes circling near the Air Force Academy are doing.

Gliders are airplanes without engines, and sailplanes are a specialized subset of gliders intended for soaring. Originally, builders did not distinguish between gliders and sailplanes. Pilots of all the earliest gliders were intent on staying up as long as possible, so they were taking baby steps toward soaring. When the sport of soaring was born in Germany in the 1920s, soaring became a recreational goal. Soaring enthusiasts soon developed a distinction between gliders and sailplanes. Technically, while all their engineless craft were gliders, only those that had a minimum sinking speed of slightly less than two miles per hour were considered to be sailplanes. In practice, people called the simpler, cheaper, less efficient ships used for initial training *gliders* and reserved the term *sailplane* for more efficient, more sophisticated craft with larger wings and higher lift-to-drag ratios.[1]

During World War II, an entirely different category of glider appeared: the troop-carrying or cargo glider.[2] Cargo gliders were not for soaring. Carrying a small group of soldiers or perhaps a small vehicle or a cannon, they were towed behind powered airplanes to the location of an airborne attack. Upon reaching the landing target, the glider released from the tow plane and landed as quickly as possible to avoid enemy fire. *Landed* is a relative term here, given that the pilot was committed to a landing not far from where he released from the tow plane, regardless of the terrain below. Most combat landings were closer to crashes than landings, and more than a few were simply crashes.

Recreational training sailplanes did not fly much like cargo gliders, and neither did the first purpose-built U. S. military training gliders (the Franklin TG-1 and the Schweizer TG-2). They were much more agile and had much flatter glides than did the cargo gliders that pilots were supposed to be learning to fly. So the U.S. military developed training gliders based on light airplanes such as Piper Cubs and Taylorcraft, removing the engines and replacing them with an extended nose containing an extra seat. In the end, the Army bought sizable numbers of both purpose-built and modified-power-plane glider trainers, but they were quickly declared surplus at the end of the war when helicopters rendered cargo gliders obsolete. The ready availability and low cost of surplus training gliders such as the TG-1 and the TG-2 fueled a postwar surge in recreational gliding in the United States.[3]

After World War II, essentially all recreational gliders were sailplanes; and over time, as memories of cargo gliders faded, the term *glider* came to be nearly synonymous with *sailplane*. Postwar sailplanes benefited from wartime advances in aeronautical sciences and became highly specialized and efficient. Sinking speed is a function of the lift-to-drag ratio; and while sailplanes from the 1930s had lift-to-drag ratios of twenty or twenty-five to one, by the 1970s, typical sailplanes had ratios of thirty to one.[4] Today's highly tuned competition sailplanes can have lift-to-drag ratios of well over forty to one.

The term *motorglider* might sound like an oxymoron, but such a craft (sometimes called a self-launching sailplane) is in fact a sailplane that incorporates a small propeller engine. Some have engines in the nose; others have engines on a strut or a pylon above and behind the pilot. Usually, a motorglider takes off using its engine and climbs to a good soaring altitude. Then the pilot shuts off the engine and either feathers the propeller blades (turns them edge-on to the airflow to reduce their drag) or, with some pylon-mounted engines, actually retracts the engine into the fuselage. Then the airplane depends on rising air, like any other sailplane, except that the pilot can restart the engine to get back to the airport if necessary.[5]

Unlike a motorglider, a typical sailplane has no way to take off on its own. How does it get up in the air to start searching for places where air rises? The earliest gliders and the first recreational soarers launched by sliding down steep hills or mountainsides to get their wings moving fast enough to produce lift. Of course, usable slopes are not available everywhere, so soaring pilots had to come up with other methods. The most effective is to use a powered airplane as a tow plane to pull the sailplane to some specified altitude, typically 2,000 feet above ground level. Then the sailplane releases from the tow plane to search for rising air.[6] Another technique is to use a powerful winch. A ground crew person stretches the towrope down the runway, with the winch at one end and the sailplane at the other. The towrope is hooked to the sailplane, and the winch starts reeling in the towrope. This quickly accelerates the sailplane to flying speed. The sailplane climbs as steeply as possible until it is almost over the winch and then releases the towrope. A third approach is to use a car or a truck in place of the winch. The tow car is connected to the sailplane by a towrope and pulls the sailplane down the runway fast enough to allow the sailplane to take off and climb. Once the sailplane gets as high as the towrope allows, it releases the towrope and goes soaring.[7] A winch tow or car tow cannot get the sailplane as high as a tow plane can, rarely higher than five hundred feet above the ground. Above valleys, however, or at times and locations with lots of rising air, these methods can be sufficient. Also, sailplanes are always at risk of off-airport landings, and sometimes a tow from a car along a convenient lane or a flat meadow can get a sailplane high enough to soar and fly back to the airport. Doing so saves the pilot and ground crew the trouble of disassembling the sailplane and hauling it back to the airport in a trailer.

Among animals, the strict gliders, such as flying squirrels and gliding lizards, don't normally have effective-enough wings to soar. They generally have to climb up a tree, glide down to a lower perch on another tree, climb up again, glide to a lower perch, and so on. The animals that are specialized for soaring are more like motorgliders: they can flap their wings to take off or to find rising air, but they much prefer to soar. These specialized soarers include vultures, eagles and large hawks, albatrosses, and frigatebirds. Other large birds, such as pelicans and cranes, and the largest species of bat soar opportunistically but readily resort to flapping when they cannot find rising air.[8] All have either moderate-aspect-ratio wings but very low wing loading so they can fly very slowly (such as vultures) or very long, high-aspect-ratio wings for very flat glides (such as albatrosses).[9]

Rising Air: Where?
Thermals

Everyone has heard the phrase "hot air rises." When something on the ground—say, a plowed field—causes air over it to warm up more than air over nearby forests or ponds, the air becomes less dense as it warms, so it rises. This rising air is called a thermal. Sometimes a thermal takes the form of a dust devil, or whirlwind. Just as a whirlpool forms over a drain as water flows out of a bathtub, a dust devil can form around a core of rising air if the source of the thermal is relatively small. Larger warm areas, such as fields, pastures, or broad sections of pavement, can produce much larger thermals, which usually take the form of huge, invisible bubbles or doughnuts of rising air. They can be any size, from a dozen meters to more than a kilometer across.[10]

Thermals tend to be strongest and most numerous during the middle of the day, when the heat of the sun is most intense. Although a solid layer of clouds can reduce or eliminate thermal activity, a sky full of separate, puffy cumulus clouds is paradoxically a sign of intense thermal activity. The rising air in each thermal cools as it rises; and with the right combination of humidity and cooling rate, at some altitude its temperature can drop below its dew point. In other words, as it rises, the air cools to the point at which it can no longer hold all its water vapor, and the water condenses and forms a cloud.

Each of those separate cumulus clouds is the top of a thermal. Sailplane pilots are usually happy to see many separate cumulus clouds because the clouds both assure them that rising air exists and make that rising air easier to find. Pilots of light powered airplanes, however, are not so fond of those conditions because they can make for a very bumpy ride.

On a clear sunny day, if the air is dry, thermals may still form, but they won't produce clouds. Because rising air is invisible, sailplane pilots depend on a sensitive rate-of-climb indicator called a variometer to tell them if they are in a thermal. Modern variometers also use an audible tone that rises or falls in pitch to indicate whether the sailplane is rising or falling in altitude so that the pilot can monitor the rate of climb while watching outside the cockpit for other sailplanes, buzzards, obstacles, and so on.[11] Sailplane pilots and soaring birds can also use other cues to find thermals. For example, movements of foliage on the ground can show that wind from several directions is converging on a point that is the center of a thermal. If a soarer gets lucky and flies through the edge of a thermal, the rising air will lift one wingtip. The soarer should then turn toward the high wingtip to enter the thermal. Soaring birds are also excellent markers of a thermal. It's not unusual to see half a dozen buzzards occupying a single thermal. Once one or two find a good thermal, others see their success and fly over to join them on the aerial escalator.

Soaring birds typically have very keen vision and, in principle, could use the apparent shrinking of objects on the ground to sense upward movement. This technique becomes less and less sensitive, however, with increasing height and probably is not very effective higher than a couple of hundred feet above the ground—not very high for a soaring bird. To the best of my knowledge, no one has studied vultures to see if they have some biological equivalent of a variometer, but I suspect that all they need are ears that are a bit more sensitive to pressure changes than mine are (see Chapter 6).

Cross-country soaring in thermals can be challenging. A soaring bird or a sailplane can gain altitude by circling in a thermal, but that does not advance the soarer much horizontally. Regions of descending air, or sink, usually occur between thermals; what goes up must come down somewhere. The trick for the cross-country soarer is to get as much height as possible in a thermal and then dash to the next one. If thermals are strong and close together, the dash should be fast and steep so that the soarer wastes less time descending. If thermals are weak and far apart, the soarer should fly slowly, at minimum sinking rate, between thermals so as to lose as little height as possible before encountering the next thermal. Paul MacCready, a champion sailplane pilot and builder of human-powered airplanes (see Chapter 10), invented a simple calculating device used with the variometer that tells the pilot how fast he should fly between thermals to maximize average horizontal speed, or distance.[12] Oddly, at about the same time as sailplane pilots were adopting this MacCready speed ring, biologists discovered that soaring birds adjust their speed between thermals in just the same way in order to maximize their horizontal distance.[13] These birds know, either instinctively or by learning from their elders, how to estimate the best glide speed between thermals; and they do it without using a speed ring or a variometer.

Slope Soaring

The other common source of rising air occurs when wind blows up a steep slope. A soarer can fly along such a slope and be held up by the upward tilt of the wind. These conditions are called ridge lift, and soaring in ridge lift is called slope soaring, or declivity soaring.[14]

The conditions producing ridge lift are rather different from those that produce thermals. Slope soaring requires a wind that is blowing directly up a slope more or less perpendicular to the ridgeline. The wind speed required to slope-soar depends on the steepness of the slope and the size of the soarer. Dragonflies can slope soar on a mild breeze blowing up a sand dune, whereas a sailplane needs a stiff wind blowing up a large hill or small mountain.[15]

One potential advantage of ridge lift is that the soarer does not have to circle in one place to gain height. If the slope is on a long ridge or a chain of hills or mountains, the soarer can climb along the length of the ridge. Some of the longest sailplane flights have been made by pilots soaring in ridge lift along mountain ranges. If the ridge happens to be oriented in the direction of the soarer's goal, so much the better. Ridge lift is, however, geographically predetermined. Where the ridge ends, the ridge lift ends as well. At the end of the ridge, the soarer will have to reverse course and go back along the ridge, find a thermal, dash to another nearby ridge, or land. Without thermals or other nearby slopes, a soarer in ridge lift is limited to tacking back and forth along the ridge.

Ridge lift comes with an extra hazard. Where the wind blows over and past the crest of the ridge, a zone of intense turbulence forms. Called rotors, these zones can be violent enough to damage sailplanes that are caught in them. Beyond a rotor, the wind blows down the downwind side of the slope, creating sink instead of lift. If a soarer accidentally flies into the turbulence at the top of a ridge, it usually gets tossed out of the rotor on the downwind slope. When that happens, the soarer is usually forced to the ground by the sinking air, regardless of how inhospitable or hazardous the terrain might be.[16]

The advantages of slope soaring more or less mirror the disadvantages of thermal soaring, and vice versa. Thermals can form almost anywhere in a haphazard pattern across the landscape, but they depend on specific weather conditions and can be challenging to find. Ridge lift occurs only where there is a ridge and whenever the wind is in the proper direction. Ridge lift is easy to find but not much use if the ridge is not aimed toward the soarer's destination. Thermals and ridge lift are geographically complementary because thermals form best over reasonably flat terrain whereas hilly or mountainous areas produce ridge lift. A soarer flying cross-country should be prepared to use both types of lift.

Soarers have discovered how to extract lift from at least a couple of other processes in very specialized conditions. For example, albatrosses use a sophisticated method called dynamic soaring to take advantage of the boundary layer of the wind over open ocean water. Albatrosses have been known to soar this way for more than 1,000 miles.[17] Although such boundary layers are too small for sailplanes to use, pilots can use updrafts caused by weather fronts as well as lee waves—a series of gigantic standing waves that form thousands of feet above and downwind from mountain crests.[18] Because these other soaring techniques are so specialized, we won't look at them any further.

Who Soars?

In a strong-enough updraft, any flying animal can soar. Insects are often carried to great heights in thermals, although with one or two exceptions, biologists don't yet know if such insects enter thermals intentionally. What we do know is that the most highly specialized soarers are all large. The largest living birds—condors and albatrosses—are soarers, as are the vast majority of birds at the large end of the size spectrum. Some extinct soarers were even bigger than condors: an extinct group of condor-like birds, the teratorns, included members with wingspans almost half again as big.[19] The largest-known flying animals, the largest pterosaurs, were even bigger than teratorns, with wingspans approaching that of a Piper Cub airplane.[20] These huge pterosaurs, as well as the teratorns, were almost certainly soarers. Few bats get as big as the larger birds, but the largest bat, the Samoan flying fox, slope-soars routinely.[21]

Big flying animals tend to soar for two reasons. First, they tend to have more efficient wings. In the size range of animal wings, as wings get bigger, even with the same general shape, their lift-to-drag ratios increase. High lift-to-drag ratios give them flatter glides and lower sinking speeds. Second, big animals tend to have less powerful muscles for their size (see Chapter 3). Because flapping flight requires a lot of power, big animals can reduce their power requirements by soaring instead of flapping. Condors and albatrosses can flap for short periods but their muscles simply cannot produce enough power to flap continuously for extended periods. Fortunately for them, their size gives them more effective wings than if they were smaller.

The best way to make a wing more efficient is to make it long and narrow—that is, to give it a high aspect ratio. Some of the soaring birds have done exactly that. Frigatebirds have aspect ratios of about ten, fulmars and pelicans of about eleven, and albatrosses of between thirteen and fifteen. Paradoxically, however, birds that soar over land have lower aspect ratios—seven for griffon vultures, turkey vultures, and storks; and only six and a half for bald eagles.[22]

Why do land birds have wings with such low aspect ratios? Soaring seabirds generally spend their time over open ocean or barren islands, whereas soaring land birds live in a much more cluttered, obstacle-strewn environment. The ten-foot wingspan of an albatross would be problematic for a bird that needs to land and take off in and around trees. Shorter wings are also easier to flap, and easier flapping lets large terrestrial soarers take off and land much more steeply than large seabirds are able to do. Albatrosses tend to roost on steep hills or cliffs to avoid long, strenuous take-off runs. In a way, steeper take-offs and landings are yet further adaptations to living around many obstacles.

Land soarers compensate for their wings' relatively low aspect ratios in a couple of ways. First, they glide with separated primary feathers (see Chapter 2), a trait that makes their wings perform as if they had higher aspect ratios because it gives them higher lift-to-drag ratios.[23] Second, they have larger wings for their size, which gives them low wing loading, typically only half that of soaring seabirds.[24] Low wing loading allows them to fly slowly, which in turn reduces their sinking speeds. A buzzard may not have as flat a glide as an albatross does; but the vulture can fly slower, so it can turn sharper to stay in thermals, and its sinking speed is low enough so that it can use even weak updrafts. The vulture might be an even better soarer if its wings were longer, but the vulture's wings are a design compromise, trading off soaring performance against take-off and obstacle-avoidance ability.

Although soaring is a virtual necessity for the biggest flying animals, at least one type of insect has evolved to take advantage of soaring.[25] Monarch butterflies migrate huge distances—up to 2,000 miles—and they routinely soar under favorable conditions. Although their wings have low aspect ratios and therefore low lift-to-drag ratios, like all butterflies they have very large wings for their size. These large wings give them very low wing loading, even for insects, so they glide very slowly. Even though they have a much steeper glide than a typical soarer does, they glide so slowly that they have a very low sinking speed. The monarch's strategy when migrating is to soar when the wind is blowing in the direction of their goal. If the wind is blowing in the right direction, monarchs circle in thermals without flapping, letting the wind push the thermal toward their destination. Monarchs, however, are more versatile than sailplanes or even motorgliders. On a day with no thermals and either light winds or tailwinds, monarchs simply fly under power like any other butterfly. A monarch in migratory conditions can flap for days without stopping to eat, only landing at night or in bad weather. Monarch butterflies thus can function as either sailplanes or efficient powered airplanes at will, something humans have yet to perfect.

Many birds also migrate enormous distances, and being moderately sized or smaller, most use powered (flapping) flight. Although a few large birds, such as Canada geese, migrate long distances entirely under power, larger avian migrants tend to soar some or all of the way. Colin Pennycuick, a leading authority on bird flight, once followed a flock of common cranes in a small airplane as they migrated across northern Europe. When thermals were present, the cranes soared. In strong thermals, they stayed between 1,500 and 5,000 feet above the ground; and in one overcast region with weak thermals, they gradually lost altitude down to about 1,000 feet. After flying into an area with stronger thermals and gaining altitude, the cranes came to a region where thermals were too weak for soaring. So when the birds drifted down to about five hundred feet, they began flapping and continued under power until they were out of Pennycuick's sight.[26]

A number of large birds, such as eagles, vultures, storks, and large hawks, migrate entirely by soaring. Storks, for instance, soar for more than 4,000 miles between sites in Europe and Africa. Because thermals don't form over water (except over the ocean in the tropical convergence zone, where frigatebirds soar), storks and other soaring migrants must soar around the ends of the Mediterranean Sea, either through the

Middle East or around the west end with a powered dash across the Strait of Gibraltar. The common buzzard, a close relative of the North American red-tailed hawk, may well be the champion soaring migrant. This hawk migrates a one-way distance of more than 6,000 miles between summering grounds in eastern Europe and western Siberia and its winter home in southern Africa. It travels exclusively by soaring and takes almost two months to travel each way.[27]

Migrating birds can use slope soaring as well as thermal soaring. Not far from Duluth, Minnesota, a large ridge near the shore of Lake Superior is oriented so that fall weather fronts make it a reliable source of ridge lift. During the fall migrations, many soaring birds— eagles, turkey vultures, falcons, hawks—collect to get a free ride and an altitude boost along this large hill, appropriately known as Hawk Ridge. The birds cannot soar over Lake Superior, so the lake shore funnels the southbound fall migrants to the ridge, which is near the west end of the lake. The birds climb up as high as possible over the ridge before setting out on the next leg of their cross-country trip. During peak days, many thousands of birds soar along Hawk Ridge, where naturalists have been counting and recording the slope-soaring birds for decades.[28] Something similar happens at other places, such as Hawk Mountain in Pennsylvania.

History of Human Soaring

Humans don't use soaring for migration nor for any practical transportation purpose. We use sailplanes almost exclusively for recreation and competition. Since the days of the earliest gliders, pilots have competed to see who could stay up the longest or fly the highest or the farthest. This competitive aspect of soaring is still going strong today.

Otto Lilienthal systematically studied flight using human-carrying gliders and in the 1890s made glides of more than 1,000 feet.[29] The Wright brothers followed in his footsteps, with Orville setting a gliding distance record of more than 2,000 feet in 1902 and a duration record of almost ten minutes in 1911.[30] His gliding distance record was not broken until just before World War I, and his duration record remained the U.S. record for almost three decades.

Soaring as a sport, separate from aeronautical research, really got its start in Germany in the 1920s. After World War I, the victorious Allies placed severe restrictions on Germany's ability to build airplanes to prevent them from reforming a strong air force. In reaction, a group of German pilots and aviation enthusiasts formed a gliding club and began flying gliders in the Rhön Mountains in 1920. They quickly discovered the power of ridge lift and began breaking records left and right. By 1922, Germans held a duration record of more than three hours and a distance record of more than three miles.[31]

By 1925, the German distance record was up to fifteen miles, but aviation enthusiasts were beginning to wonder if tacking back and forth across the same slope for hours was a serious challenge.[32] Then in 1926, Max Kegel breathed new life into soaring by accidentally flying into a thunderstorm. The storm's updrafts blew him so high that he landed thirty-four miles away from his launch point, and thermal soaring was born. In 1929, Robert Kronfeld won a 1,200-dollar long-distance prize by flying ninety miles, mostly by using thermals under and within clouds.[33]

The Germans built up such a huge lead in distance records in the 1920s that it is amazing that U.S. pilots were able to catch up. For instance, in 1932, the German record soaring distance was 143 miles, but the U.S. record was only 66 miles. In 1934, however, American Richard du Pont set a new world record of 158 miles. Alas, this record did not last long. The next year, three Germans flying in a competition shattered the old mark by soaring 313 miles, almost double du Pont's distance.[34] In 1939, Russian Olga Klepinkova, became the first woman to hold a world soaring distance record with a distance of 465 miles.[35] After World War II, U.S. pilots once again became competitive in soaring distances, holding three world records into the 1970s. In 1972, Hans Grosse beat the American record of 717 miles with a 908-mile flight, and the Europeans have dominated the world soaring distance records ever since.[36] In fact, in 2006 the U.S. distance record, set back in 1984, was fives miles shorter than Grosse's 1972 world record and barely half the current world record.[37]

The early German pilots worked hard to set duration records as well as distance records. In 1920, they could not even equal Orville

Wright's 1911 mark of 9¾ minutes. By 1921, however, several pilots had beaten Wright's time; and by 1922, with the first true sailplanes, pilots were making flights lasting from one to three hours.[38] After a brief challenge by French pilots Alphonse Thoret and Alexis Maneyrol in 1923, the Germans reclaimed the duration record in 1925 with flights of more than twelve hours.[39] Cruising back and forth along the same ridge for a whole day may have done little more than prove, as the wags put it, that the pilot's rump was as hard as his head. But as Daniel Halacy points out in *With Wings As Eagles*, those pilots could not do anything else because they had not yet discovered thermal soaring.[40]

After Max Kegel accidentally discovered updrafts in and near clouds, duration records continued to climb. In 1931, an American, William Cocke, soared over the Hawaiian islands of Oahu for more than twenty-one hours; and in 1948, a Frenchman, Guy Marchad, nearly doubled that time with a flight of slightly less than forty-one hours.[41] In 1952, Charles Atger, another Frenchman, raised the bar to an astonishing fifty-two hours.[42] Although other solo pilots have tried to beat his time, and more than one has died in the attempt, only teams of pilots in two-seat sailplanes have ever done so. After one such team in 1961 raised the record to more than seventy-one hours (again over Oahu), the international body that sanctions aviation records—the Fédération Aéronautique Internationale (FAI)—decided that endurance records had become too dangerous and retired the duration category.[43]

The other two main categories of records in soaring are altitude and speed. Progress in these categories initially paralleled progress in distance and duration, and that is still largely true for cross-country speed records. Altitude records are a different story. In the 1920s, German pilots had pushed the altitude record to more than 8,000 feet; and in 1934, Heinrich Dittmar, a German pilot flying in Brazil, soared in clouds to more than 14,000 feet.[44] In 1937, "a Russian named Fydaroff" more than doubled this record by reaching 39,950 feet, but his new record was controversial because he was towed to 28,000 feet before he started soaring.[45] After World War II, pilots in the

United States and New Zealand discovered mountain waves, fantastically immense standing waves of wind that form many thousands of feet above high mountain ranges. These standing waves have zones of strong lift as well as zones of strong sink. In 1950, U.S. pilot Bill Ivans soared to more than 42,000 feet over the Sierra Mountains in California.[46] In 1954, British pilot Philip Wills set a British record of more than 30,000 feet when he stumbled into wave lift during what started as a routine flight in New Zealand.[47] In 1961, Paul Bikle, a NASA researcher, reached more than 46,000 feet in waves over the Sierras. Although his absolute altitude record has been broken, his gain-in-altitude record stands to this day. He started from about 4,000 feet above sea level (about 1,500 feet above the ground) and gained more than 42,000 feet.[48] Another American, Robert Harris, also flying in wave lift over California, set a world absolute-altitude record of more than 49,000 feet in 1986 that stood for twenty years. That altitude is more than nine miles high, and Harris achieved more than seven and a half miles of that climb by soaring. The current absolute-altitude record of just over 50,000 feet was set by American adventurer Steve Fossett in 2006.[49]

To put these altitude records into perspective, consider that typical piston-engine airliners from about 1960 had service ceilings of approximately 25,000 feet above sea level. Today's modern jetliners rarely fly above 35,000 feet and typically have service ceilings of about 45,000 feet.[50] These airliners have pressurized, heated cabins. Yet sailplane pilots in the 1950s and 1960s were soaring unheated, unpressurized sailplanes to altitudes of more than 40,000 feet. Conditions at these altitudes are frighteningly harsh. At 42,000 feet, Bill Ivans recorded the air temperature to be sixty-nine degrees below zero Fahrenheit; and upon reaching 30,000 feet, Philip Wills decided he'd better descend when his canopy started to crack from the cold.[51] Oxygen is an even bigger concern. Air pressure decreases with altitude, so less oxygen is available the higher you go. Most people experience the first minor symptoms of oxygen starvation at 10,000 feet, and the average person risks unconsciousness when breathing unpressurized air at 17,000 feet or above. Paul Bikle flew more than twice that high, and Robert Harris flew almost three times that high. Pressurized cockpits and heaters are far too heavy and bulky to fit into a high-performance sailplane, so high-altitude sailplane pilots must dress warmly and wear a mask delivering pure oxygen from a small compressed oxygen tank. Indeed, Paul Bikle cut short his record climb while still in strong lift because he was not sure his oxygen would last long enough for him to descend to breathable air. The obstacles to soaring at very high altitudes are clearly formidable, which is why these records have stood for so long. Until someone develops a high-altitude pressure suit light, compact, and flexible enough to wear in a sailplane as well as lighter-weight, longer-duration oxygen systems, breaking the absolute-altitude and altitude-gain records will be nearly impossible.

Unlike the altitude records, soaring pilots have set new distance records regularly; and in recent years, none have been held by U.S. pilots.[52] For instance, the current free distance world record (distance traveled with no prearranged checkpoints or goal) of 1,363 miles was set by New Zealander Terrence Delore in 2004. The top U.S. free distance mark is much shorter, at only 903 miles (set in 1984). Or consider the more exacting task of the distance to a predetermined goal with fixed checkpoints. German pilot Klaus Ohlmann set a world record of 1,319 miles in 2003, also beating the U.S. record of 801 miles (set in 1977) by a wide margin. With duration records discontinued and altitude records limited by severe environmental challenges, distance records are still vulnerable to a combination of piloting skill, incremental improvements in sailplane performance, and favorable atmospheric conditions. Speed records, which the FAI keeps over distances ranging from 100 kilometers (61 miles) to 1,500 kilometers, are similarly vulnerable: pilots have set eight of the eleven world records since 2003.[53]

Why Aren't Birds Better?

The specialized soaring birds have the highest lift-to-drag ratio in the animal kingdom. Large hawks and eagles have lift-to-drag ratios of about ten, turkey vultures of about sixteen, and wandering albatrosses of about nineteen. Although these values are extreme for animals, they are not even in the same league with sailplanes. Even early sailplanes from the 1930s had lift-to-drag ratios of twenty to twenty-six; and over the past few decades, the lift-to-drag ratio of the most advanced sailplanes has continued to climb.[54] Nowadays, two-seat basic sailplanes used for training have lift-to-drag ratios of twenty-three or higher, and advanced competition machines such as the Schleicher ASW 22 and the Schempp-Hirth Nimbus 4 have lift-to-drag ratios of fifty-eight to sixty.[55]

Why are sailplanes so much more aerodynamically efficient than birds? Looking at a sailplane's wing alongside a bird's wing sheds some light on this disparity. Sailplanes get their high lift-to-drag ratios mainly from two features: extreme aspect ratios and specialized, precisely formed airfoils. Both the ASW 22 and the Nimbus 4 have aspect ratios of about thirty-eight, meaning that each wing is about *forty times* longer than it is wide. Moreover, the wings are not just skinny; they are long: eighty-two feet for the ASW and eighty-six feet for the Nimbus 4. That is longer than the span of a small jet airliner. Off-airport landings being an occupational hazard in soaring, most sailplanes have detachable wings so that each half can fit into the recovery trailer along with the fuselage. The most advanced sailplanes take this a step further. Their wings are so long that when the ground crew disassembles one to recover it from an off-airport landing, the wings come apart into four (ASW 22) or six (Nimbus 4) sections to fit into the recovery trailer. The only way to make such immensely long, skinny, yet precisely shaped wings is with high-strength composites, typically graphite-epoxy (see Chapter 2).[56]

The airfoil shape is also important. Sailplane designers and aeronautical engineers have developed special airfoils for sailplane wings that have high maximum lift-to-drag ratios but also maintain high lift-to-drag ratios over a wide range of speeds.[57] Small variations away from the desired airfoil shape can cause disproportionately large reductions in the wing's performance, so designers need smooth, precisely sculpted wings to realize the potential of these special airfoils. Again, only molded composites or composites formed over Styrofoam cores can give wings with accurate-enough airfoil shapes.

Sailplanes are the only aircraft outside of the homebuilt category in which designers have used fiberglass extensively. In the 1950s, European sailplane manufacturers discovered that they could make very smooth, flowing, seamless shapes from fiberglass, which so reduced the drag and improved the lift-to-drag ratio that they were willing to accept a slight weight penalty.[58] Modern composites are now much better than fiberglass: not only are wings and fuselages lighter, but they are also stronger than metal.

Now consider the construction of a bird's wing: skin, muscle, and bone, the same as any warm-blooded animal, plus feathers made from protein. These materials simply do not have the strength necessary to make an extremely long, extremely narrow wing. Yet even if they could be formed into such a wing, would it do the bird any good? Bird wings, even those of vultures and albatrosses, have to be flapped because they have to at least occasionally act as propellers as well as wings (see Chapter 3). I doubt if a bird could flap a wing with an aspect ratio of thirty-eight, and how would it ever fold up such a long wing on the ground? The need to flap also means that the wing needs to be inherently flexible (see Chapter 2), so a bird's wing could never maintain a precise airfoil shape that accurately emulates a sailplane's airfoil.

True, a buzzard cannot match the aerodynamic performance of a typical sailplane, but consider what the buzzard *can* do. It weighs less and flies more slowly than the sailplane, so it can use weaker thermals than any but perhaps the most advanced sailplanes and can turn more sharply to stay in much smaller thermals. If the buzzard can't find thermals or ridge lift, it can switch from soaring to flapping (powered flight) to get home. It can take off almost vertically. It can land on a tree branch and even fly through trees if they are not too close together. After it lands, it can fold up its five-foot-long wings so compactly that a person who has never seen a bird might not realize the buzzard had wings at all. Taken together, the buzzard is a near equal to the sailplane in aerial ability and far and away more versatile. Yes, the sailplane is aerodynamically more efficient; but if I had to fly for a living, I think I would prefer the buzzard's versatility.

~ 8 ~

Straight Up
Vertical Take-Offs and Hovering

Several times each year, I pass through the Kansas City International Airport. It is an enormous facility with three runways, each about two miles long, located well outside the urban sprawl of Kansas City. The airport is so big—about fifteen square miles—that I spend almost ten minutes on the freeway when I drive around roughly half of the airport's perimeter. Yet as big as it is, the Kansas City airport is far from the largest airport in the world. Hartsfield-Jackson International Airport in Atlanta has five runways, Chicago's O'Hare International Airport has six runways and is considering adding two more, and Denver's airport covers almost three times the area of Kansas City's.

Runways are the largest, most quintessential feature of any airport, and 9,000- to 12,000-foot runways are typical at airports serving passenger airliners. From my seat in the Kansas City terminal, I can see that fully loaded jetliners use more than half of the runway to take off and a bit less than half to land. Airplanes need these long take-off and landing rolls to accelerate to and decelerate from flying speeds—typically 70 miles per hour for a small propeller-driven trainer up to 180 miles per hour for a loaded jetliner. Without long, straight, flat runways, conventional airplanes cannot fly.

Now contrast an airplane with a flock of pigeons pecking for crumbs in a city park. If I suddenly clap my hands, the pigeons rise up vertically to a height well above my head before flying away from me. Most flying animals can take off and land vertically, but conventional fixed-wing airplanes cannot. If we need to take off vertically or hover, we use more complex, specialized machines such as helicopters.

Hovering and vertical take-off or landing are not identical actions, although they are closely related. For all practical purposes, a flying machine must be able to hover (that is, fly with no ground speed and little or no airspeed) in order to land or take off vertically, and often such a machine's hovering ability is as useful as its ability to land in a tight spot. Flying animals, in contrast, show more of a distinction. All insects and bats and most birds can take off at a steep angle with little or no ground run. Hovering, however, requires extra specializations and high power per unit weight. Thus, only some insects, bats, and small birds routinely hover for more than a couple of wing beats. For our purposes, hovering implies vertical take-off (and vice versa) for machines, but many animals may be able to take off vertically yet have little hovering ability.

Why Can't Everyone Hover?

Hovering and its close relative, vertical take-off and landing (VTOL), happen with no forward speed. This condition immediately causes a problem for a flyer: wings only produce lift when they move through the air (see Chapter 2). So if I want to fly with no forward speed, I need to come up with some way for my wings to move while the rest of me does not. If I want to hover with wings, in other words, then I need to put power into moving the wings even when I am not moving through the air.

Power, or the rate of doing work, is equal to force times speed (discussed in Chapter 3). Because flight is usually much faster than other kinds of locomotion, these high speeds mean that flight in general requires a lot of power. Paradoxically, hovering, or flight with no speed, is the most power-hungry form of flight. The problem is that, when hovering, a flyer cannot take full advantage of effective wings. That is, the flyer cannot produce a little thrust to move the wings through the air and get a lot of lift in return. Instead, a hovering flyer must use brute strength to move its wings directly and fast enough to produce the lift needed to support its weight. So even though a flyer needs a great deal of power to fly and even more power to fly fast, it needs the most power of all to hover.

Hovering also brings new control problems. In normal forward flight, flyers turn by banking and, for efficient flight, need to keep facing into the relative wind produced by their movement (see Chapter 4). Airplanes also get significant control and stabilizing action from their tails, although this ability is less important for flying animals (see Chapter 5). But none of these processes work in hovering. Banking merely causes a hovering flyer to fly sideways, and tail control surfaces are useless without air flowing over them. Hovering flight requires completely different methods for maneuvering.

Once you figure out how to maneuver while hovering, however, the payoff is huge. Flight to the side, straight up or down, or even backward is possible. So is rotating in place, in which the hoverer faces a new direction without any linear movement. The tricks to control and stabilize a hover require relatively minor modifications for a flapping flyer but are extremely challenging for engine-driven machines.

Most flying animals can take off vertically, and of those, all but the largest can hover at least briefly. Size plays a very different role for hovering animals than it does for soarers or flying machines. Among flying animals, it pays to be small if you want to hover. But if wings become less effective as they get very small (discussed in Chapter 2), how does small size benefit hovering? The key is the high power requirements of hovering. Ounce for ounce, small animals tend to have more powerful muscles than large animals do.[1] Thus, with flight muscles making up the same proportion of its body mass, the specific power output (power per unit weight) of a small flyer will be higher than will the specific power output of a big flying animal (see Chapter 3). In effect, an ounce of hummingbird muscle is more powerful than an ounce of eagle muscle.

Small flyers have other advantages as well. As animals get smaller, their weight decreases much faster than their linear dimensions because weight is proportional to length cubed. For example, imagine a large goose with a wingspan of sixty inches and a small duck with a wingspan of thirty inches. Their bodies are the same general shape, but the goose is about twice the length of the duck. Therefore, its weight should be two cubed, or eight times greater than the duck's. If the goose weighs thirteen pounds, we would predict the duck's weight to be 1.6 pounds—that is, thirteen pounds divided by eight. (Because their bodies are not exactly geometrically similar, the duck probably weighs closer to two pounds, still a vast drop in weight for a modest decrease in length.) The upshot is that smaller flying animals have proportionately lighter wings with larger surface area for their size. Not only are these small wings easier to flap (less inertia), but they also have lower wingloading—more surface area for their weight. This is why all insects can potentially hover; and for small ones, such as fruit flies, hovering is probably not much harder than forward flight. Even large insects such as dragonflies and bumblebees can hover for extended periods.

Now let's consider vertebrates. When people think of hovering animals, most probably think of hummingbirds. Indeed, hummingbirds are both the smallest birds and the ones that are best at hovering. Other small birds such as wrens and finches can hover readily and often do so when approaching a perch or a nest entrance, but they are not nearly as adept as hummingbirds and probably don't have the power reserves to hover for more than a few seconds. Larger birds up to pigeon or crow size can hover for a few wing beats, perhaps a second or two, and can also take off vertically. In fact, some birds as big as ducks can take off vertically when startled, although this is an extreme maneuver; most birds larger than crows prefer an initial flight path that is more horizontal, which allows them to build up flight speed gradually. When taking off from a level surface, birds the size of geese or larger usually require a lengthy horizontal run to accelerate to flying speed, not unlike an airplane. Such large birds get no closer to hovering than to make a couple of wing flaps to brake just before touchdown.

For a flying machine, hovering and VTOL require huge specializations and a design concept completely different from that of conventional airplanes. Fixed wings are useless; something else needs to replace this fundamental airplane component. The most successful approach is to replace the wing with a rotor. A rotor, somewhat like a giant propeller on a vertical shaft, actually combines the functions of a propeller and a wing. (This, not coincidentally, is exactly what the flapping wings of flying animals do, and several biologists have noted that animal wings have more in common with helicopter rotors than with fixed airplane wings.)[2] Rotors must be actively turned by an engine, which brings us back to the power issue. Because of the high lift-to-drag ratio of wings, a fixed-wing airplane's engine needs to produce only enough thrust to overcome the airplane's drag, typically about one-tenth of lift produced by the wings, which supports the weight. For hovering, however, the engine must produce enough thrust to support the craft's entire weight, requiring a great deal more engine power. This huge power requirement demands a proportionately huge fuel consumption, which in turn means that all time spent hovering puts a huge dent in horizontal flight range. Because the power required to hover is directly related to weight, weight reduction is even more precious for hoverers such as helicopters than it is for airplanes.

Helicopters (aircraft that fly using one or more powered rotors) have some inherent control problems. First, in a hover, a tail is useless as a stabilizer; so right off the bat, helicopters lose a major source of an airplane's stability. Second, helicopters with a single rotor suffer from the torque reaction: according to Newton's third law of action and reaction, turning the rotor in one direction produces a reaction that causes the rest of the craft to turn in the opposite direction. In other words, if we imagine that we're looking at a helicopter from above, we can see that if the rotor turns counterclockwise, the fuselage will turn clockwise unless something prevents it from doing so.[3] Moreover, if the rotor blades are rigidly attached to their shaft, as propeller blades are, any gust that tilts the craft will also tilt the rotor, causing it to drift in the direction of tilt. This drift usually causes the craft to then tilt in a new direction, eventually leading to wild swings and oscillations.[4] Overcoming these control issues as well as hoverers' power-versus-weight limitations has been challenging; experimenters continued to work for four decades after the advent of the first successful airplane before they were able to produce a successful helicopter.

How to Hover

Just as a bird in forward flight has fundamental similarities with and differences from a flying airplane, a hovering bee or a hummingbird shares certain features with a hovering helicopter but also has major differences. How do flyers hover? Let's begin by looking at the basic operation of a helicopter.

Helicopters

A helicopter rotor is essentially a long, narrow wing that rotates about a central shaft or mast. Each blade has a typical airfoil shape and produces lift in the same way that a wing does; it simply travels in tight circles rather than a straight line. When a helicopter hovers, the rotor turns in a horizontal plane with the shaft vertical, and the lift it produces is straight up (fig. 8.1). This arrangement is fine for hovering or flying straight up and down, but what if we want to use the helicopter to actually go somewhere?

The basic controls of a helicopter, which are somewhat analogous to an airplane's aileron and elevator, are the cyclic and collective pitch controls. The cyclic pitch control is like an airplane's control column, a joystick mounted right in front of the pilot, while the collective pitch control is a lever that moves vertically on the pilot's left side. Each blade of the rotor is attached to the mast by a pivot that allows the blade to change its angle of attack, or pitch. The cyclic control makes each blade increase its pitch during a particular part of its rotation and decrease it during the other parts. When the pilot moves the cyclic control, each blade's pitch cycles up and down as it rotates—hence, the term *cyclic pitch*.[5] The cyclic control causes the rotor's plane of rotation to tilt, which tilts the direction of lift and in turn causes the helicopter to move horizontally (fig. 8.1).

But the way in which these changes in angle of attack cause the rotor to tilt is very non-intuitive. We might expect, for example, that

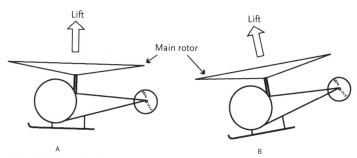

Figure 8.1   Net lift on a helicopter rotor (side view). On the left (A), the rotor is horizontal, so lift is straight up for hovering. On the right (B), lift is tilted forward, so the helicopter flies forward.

when the pilot pushes the cyclic stick forward to fly forward, the rotor blades will have a low angle of attack at the front of their rotation and a high angle of attack at the back. According to this assumption, with less lift at the front and more at the back, the rotor's plane of rotation will tilt forward so that the lift it produces is also tilted forward. In fact, however, a helicopter rotor acts as a gyroscope; and to tilt a gyroscope, you have to push on it at right angles to the direction in which you want it to tilt. For example, to tilt forward a horizontal gyroscope (spinning about a vertical axis), you have to push up on one side rather than the back. So to tilt a helicopter rotor forward, the cyclic control actually causes the pitch to increase on one side and decrease on the other (the choice of side depends on which direction the rotor is rotating) rather than at the back and front.

Helicopter builders call this phenomenon the "ninety-degree phase lag."[6] The high angle of attack on one side pushes up on that side, so the rotor tilts forward. Now the lift is tilted forward, and the helicopter will fly forward (fig. 8.1).[7] If the pilot pushes the cyclic stick to the left, the angle of attack of each rotor blade changes at the front and back of its rotation so that the plane of rotation of the rotor tilts left. If the helicopter is hovering with the rotor plane tilted left, it drifts sideways to the left. It the helicopter is flying forward, the tilted rotor acts like an aileron and banks the helicopter to the left, producing a left turn as it would in an airplane.

The other primary control is the collective pitch control lever. This control increases or decreases the average angle of attack of the rotor blades throughout their rotation. Increasing the collective pitch raises the blades to a higher angle of attack, which increases total lift; decreasing the collective decreases the total lift. So the collective pitch control affects vertical speed. The pilot can use it to make a hovering helicopter fly straight up or down and to make climbs or descents in forward flight. The collective pitch control also includes the engine throttle, usually as a motorcycle-style twist grip. This is a logical combination because increasing or decreasing the collective pitch will change the power required from the engine, so the two controls need to be adjusted together.[8]

What about the torque reaction in which the fuselage tends to rotate in a direction opposite from the rotor blades'? Helicopter designers have figured out two practical solutions. One is to use two main rotors turning in opposite directions. The torque reactions of the two rotors thus cancel each other. But two rotors greatly increase the weight and complexity of the drive train and so are rare except for large helicopters. By far the most common solution is to use a single main rotor with an anti-torque tail rotor. The tail rotor is really just a propeller that faces sideways and pushes sideways on the tail to counter the torque turning tendency.

Looking down on a helicopter from above, we can see that if the main rotor turns counterclockwise, the fuselage tends to turn clockwise; so the tail rotor needs to push to the right to counter torque.[9] Yet a tail rotor does more than merely counter torque. Its blades have adjustable pitch, just like an adjustable pitch propeller or the collective pitch of the main rotor. Increasing the pitch (or angle of attack) of the tail rotor causes its blades to take a bigger bite of the air and push harder on the tail; lowering the pitch weakens its sideways push and allows the torque reaction to take over. This gives the pilot a way to make the craft yaw left or right. So the tail rotor pitch control is connected to pedals, which the pilot uses exactly like airplane rudder pedals to swing the nose of the helicopter right or left. The big difference, of course, is that helicopters can yaw right or left even when they are not moving forward.[10]

Animal Hovering

The general idea behind a helicopter rotor is fairly intuitive: take a big propeller and point it straight up to get an upward force. Getting a similar force using flapping animal wings is less obvious. How do flying animals hover?

Animals can hover by flapping their wings in several different ways. One of these methods, horizontal stroke plane hovering, is fairly easy to understand. Because biologists figured out this method first, they also call it conventional hovering. In conventional hovering, the animal adjusts its wing beats so that its wings flap more or less horizontally (hence, "horizontal stroke plane") rather than up and down.[11] Most animals do this by tilting their bodies almost vertically, head up. With the wings moving horizontally, they can now produce lift on both the upstroke and the downstroke. During the downstroke, the wing moves forward with the normal top surface up. At the beginning of the upstroke, however, the animal flips the wing over, so that the bottom surface now faces up, and swings the wing backward. At the end of the upstroke, the animal flips the wing back upright and is ready to make the next downstroke (fig. 8.2). Lift is mostly upward on both the downstroke and the upstroke. Any fore-and-aft tilt during one half-stroke is for the most part canceled by an opposite tilt during the next half-stroke.

Scientists have studied conventional hovering in hummingbirds, large nocturnal moths called hawk moths, and bees and believe that most animals hover using some variation of this method. Ideally, the downstroke and upstroke should be perfect mirror images, but we now know that animals do not use such symmetrical strokes. Hummingbirds come fairly close, but even they produce significantly more lift during the downstroke and less than half during the upstroke.

A number of insects hover with their bodies horizontal rather than vertical and with their wings flapping more vertically than horizontally. Dragonflies are the best known of these insects. You may have

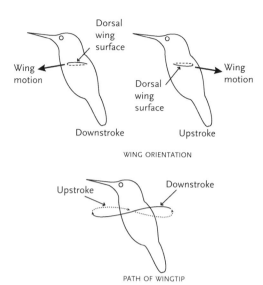

Dorsal wing surface

Wing motion

Downstroke

Dorsal wing surface

Wing motion

Upstroke

WING ORIENTATION

Upstroke

Downstroke

PATH OF WINGTIP

Figure 8.2 A hovering hummingbird moves its wings horizontally. *Top:* During the downstroke, the wing sweeps forward. At the beginning of the upstroke, the wing flips upside down and sweeps backward. (The wing airfoil is shown with the dorsal, or anatomical top, surface as a solid line and the lower surface as a dashed one.) Note that the wing is flipped upside down for the upstroke. *Bottom:* In conventional hovering, the wingtip traces a flattened, horizontal, figure-eight path. (The downstroke is shown as solid line, the upstroke as a dotted one.)

seen a dragonfly pausing to hover horizontally as it patrols the shoreline of a pond. Flies in the family Syrphidae get their common name *hover flies* from their habit of hovering in one spot to watch for prey or possible mates; and they, too, hover with the body horizontal. (Many species of syrphid flies mimic bees, so if you see a "bee" hovering with the head straight forward rather than tilted noticeably up, you are almost certainly looking at a fly rather than a bee.)

Because these insects hover with their wings beating as much up and down as back and forth, biologists have dubbed this technique "inclined stroke plane" hovering.[12] The aerodynamic mechanism has puzzled researchers for decades, but some recent computer models suggest that these flyers may actually support a significant amount of their weight at certain phases of the stroke using drag rather than lift. In other words, a dragonfly apparently turns its wing broadside into the wing motion as it flaps down. The wing, though stalled, produces

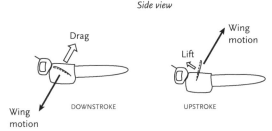

Figure 8.3 A hovering dragonfly supports most of its weight with drag produced on the downstroke; the backward tilt is canceled by lift directed slightly forward on the upstroke. During the upstroke, the wing is tilted dramatically leading-edge-up (dorsal surface shown as a solid line), and the wing operates at a negative angle of attack, producing lift on the lower surface.

a substantial drag force. Because it is moving mostly down, the drag force is mostly up (fig. 8.3). The slight backward tilt of the drag during the downstroke is counteracted by a small amount of forward force on the upstroke, during which the wing operates more conventionally to produce mostly lift and little drag.

If the computer simulations are correct, a couple of different factors are at work. First, because of their size, insect wings tend to have low lift-to-drag ratios (see Chapter 2). This means that when they operate their wings as normal lift-producing airfoils at high angles of attack for extra lift (as during hovering), their wings will also have rather high drag. Jane Wang, a Cornell University physicist, showed that using the wing as a paddle (producing mostly drag) rather than as a conventional, lift-producing wing means that an insect can produce as much or more weight-supporting force upward with approximately the same power output. Wang noted that inclined stroke plane hovering costs little more (and may even cost less in some situations) than conventional hovering does.[13] Dragonflies, for instance, are visually oriented predators, as are hover flies. Perhaps inclined stroke plane hovering is an advantage to keep them facing forward when watching for prey.

As far as hovering goes, the big difference between flying animals and flying machines is that animal hovering requires strong flight muscles and exaggeration of some aspects of the slow-flight wing-beat pattern for flapping flyers. Machines, however, require a completely different approach, using a fundamentally different mechanism from conventional flight.

For a flying animal, hovering is just an extreme form of what it already does. We can view animal flight as a continuum in both power output and wing-beat complexity: from gliding, to fast forward flight, to slow flight, to hovering. On this continuum, hovering is at one end—admittedly, the strenuous end—of the spectrum, quantitatively but not qualitatively different from forward flight. In fact, the wing-beat pattern for hovering is usually simply an exaggeration of the wing-beat adjustments a flying animal uses to fly slowly. This situation is exactly reversed in the airplane case. Fixed-wing airplanes have no way to get lift when standing still, so we have to use a completely different kind of craft, such as a helicopter, if our mission involves hovering.

What's So Great about Taking Off Vertically?

For both animals and flying machines, the ability to land and take off vertically opens up many new opportunities. For nature's flyers, VTOL gives them the ability to perch and nest in complex, cluttered habitats such as forests and thickets. They can also nest in tight quarters such as tree holes or small caves and deep shafts such as chimneys and hollow tree trunks. Animals that need significant horizontal take-off or landing runs are excluded from such habitats.

Extending VTOL to true hovering adds at least one further niche possibility: pollination. Pollinating birds, bats, and insects are all adept at hovering. Hovering lets a pollinator lap up nectar from a flower too delicate to support the animal's weight.[14] Hovering is also handy for quickly probing and sipping from many small, closely spaced flowers. Hummingbirds rarely alight on flowers they drink from, and they are masters at matching their movements to the movement of flowers that are bobbing and weaving in puffs of wind. Pollinating bats have not been studied nearly as extensively, but they are also superior hoverers. Although bees often land on flowers they visit, most are capable of drinking nectar while hovering if the flowers are small enough and the bees are thirsty enough.

The benefits of VTOL for flying machines are broadly similar to those for animals: VTOL allows them to operate without runways. For instance, an urban area does not have room for runways because it is cluttered with tall buildings, towers, and so on. As a way to deal with this problem, a number of companies have attempted to operate passenger helicopters from urban areas, usually as a quick way to get from the city center to an outlying airport or from one congested city center to another relatively close one. None of these operations has been able to stay in business, however, due to helicopters' high operating costs and public safety concerns.[15]

Yet helicopters have become common in more specialized arenas, most frequently in civilian life as life-flight ambulances. A helicopter can land directly at the scene of an accident or a disaster and rapidly transport gravely injured victims to a hospital. Due to their high operating costs and the lack of suitable landing spots in cities, helicopters cannot replace conventional ambulances in built-up urban areas. But in rural and suburban areas far from hospitals, life-flight helicopters are lifesavers; and most U.S. hospitals now have helicopter landing pads on their roofs or in their parking lots.[16]

Similarly, helicopters are extremely useful for rescuing people in otherwise inaccessible locations. Indeed, almost as soon as Sikorsky's first production helicopter, the YR-4, entered service near the end of World War II, one was involved in a spectacular rescue in Burma. In April 1944, a small Allied plane carrying wounded soldiers crashed in the jungle behind Japanese lines. The small, flimsy, fabric-covered YR-4 managed to pick up the soldiers, one at a time, and fly them to safety.[17] The U.S. Coast Guard has been using helicopters for decades to rescue people from sinking boats, and the military uses them to rescue downed pilots and evacuate wounded soldiers. Nepal operates very specialized helicopters, flown by highly skilled pilots, to rescue sick and injured climbers from the slopes of Mount Everest and other Himalayan mountains. These Nepalese pilots fly well above the service ceilings of most helicopters.[18]

Offshore oil rig operators often use helicopters. Many oil rigs are so far from shore that a boat can take the better part of a day or more to reach them. But a helicopter, flying ten to fifteen times faster than a boat, reaches rigs much more quickly and can land right on an oil rig's platform. In this situation, the expense of helicopter operations is more than outweighed by the time saved in transporting people and lightweight equipment and supplies.

For the military, of course, VTOL aircraft are worth the expense in several situations. Helicopters can operate without runways, so they can move troops and supplies directly from rear assembly areas or supply dumps to near the front lines. Indeed, the venerable Huey helicopter (officially the UH-1 Iroquois) is probably the single most evocative and enduring symbol of the war in Vietnam. Since their pioneering uses in the Korean War, helicopters have become vital for a wide array of transport and rescue functions in the modern military.

The U.S. Army and other large militaries have also been drawn to the idea of attack aircraft that can operate without runways. At first, militaries simply added machineguns or rockets to utility helicopters such as the Huey, but they eventually concluded that dedicated attack helicopters would be more effective. The U.S. Army has opted for small, agile, highly specialized, two-seat attack choppers such as the AH-1 Cobra and the AH-64 Apache.[19] In contrast, the Soviet Union developed the large, powerful, but less-than-agile Mi-24 Hind attack helicopter, which in addition to copious weapons can carry a squad of soldiers.[20] Although a VTOL attack craft sounds like a formidable concept, helicopters are not ideal in this role. Compared to other combat aircraft, they fly more slowly and fight at lower altitudes and are very vulnerable to enemy ground fire.

The real lure of VTOL flight must surely be the idea of flight from anywhere by anyone. Several companies produced successful helicopters at the end of World War II and shortly thereafter. Given the pace of technological advance at that time, several leaders in the aircraft industry touted the idea of "a helicopter in every garage." Lawrence D. Bell of Bell Aircraft Company and Igor Sikorsky of Sikorsky Aircraft both used the phrase, assuming that they would enter a booming civilian market after the end of World War II.[21] They predicted that technological progress would make piloting a helicopter as easy as driving a car and that technology and volume production would reduce the price and operating costs of small helicopters. The lure, of course, was imagining a pilot's ability to take off from her driveway and fly anywhere she wanted to without having to follow indirect or winding roads, wait for stoplights, sit in traffic jams, and so on.

So why aren't we flying our own personal helicopters to work today? First, while technological advances have simplified helicopters to a certain extent, they are still far more complex, delicate, expensive, and maintenance-hungry than an automobile (or even most airplanes, for that matter). Second, flying helicopters has not really gotten much easier: piloting a helicopter is still considered to be one of the most challenging flying tasks, mainly because a change in any one control—cyclic, collective or tail rotor pitch, or throttle—tends to affect other controls. (Learning to hover a helicopter has been compared to learning to juggle while standing on a beach ball.) Finally, even if computers could reduce the difficulty of flying a chopper, the image of thousands of people in every suburb taking off each morning and flying away willy-nilly is terrifying. The current number of accidents on our streets and highways would surely be dwarfed by the aerial carnage created by a large, unconstrained population of aerial commuters in helicopters.

Why So Long to Hover?

We don't know when the first flying animal evolved the ability to take off vertically or hover, but it was surely tens, if not hundreds, of millions of years ago. Insects were the first animals to evolve flight (more than 300 million years ago), and we have seen that vertical take-offs are easier for tiny flyers than for big ones and that flapping flight lends itself to hovering. Ancestral dragonflies were probably hovering before the rise of the dinosaurs.

But technological flight evolution has followed a very different route from Mother Nature's. Because conventional airplanes are fundamentally incapable of hovering, vertical flight and hovering required a completely new type of craft. Although helicopter-like machines made feeble, semi-controlled hops in the same decade that the Wrights first flew (and toy helicopters had been around for decades longer), no really successful helicopter flew until the 1930s, and truly practical helicopters did not come on the scene until the end of World War II. Why did inventors need nearly four decades after the success of the Wright brothers to perfect the helicopter?

Toy helicopters that could fly vertically had been around since the early nineteenth century, and Wilbur and Orville Wright even mention a rubber band–powered toy helicopter as one of their early inspirations.[22] Indeed, the term *helicopter* was coined by Vicomte de Ponton d'Amecourt in 1863 to describe his small steam- and clockwork-powered models; clearly, the concept of vertical flight using rotors is a venerable one.[23] A number of inventors built machines intended to hover and take off vertically between 1907 and the mid-1920s. Frenchman Paul Cornu made some brief hops in a tandem-rotor machine in 1907, but his engine was not powerful enough to sustain a hover. That was probably just as well because his machine would probably not have been controllable. Even so, Cornu is often listed as the first to fly in a helicopter.[24]

Engine power limits dogged would-be helicopter builders into the 1920s, when engines finally became light yet powerful enough to achieve hovering. At that point, helicopter researchers discovered that power was only the first hurdle. Hovering and transitioning from a hover to forward flight required much more control sophistication than they had expected. Working in Europe, Argentine-born engineer Raul Pescara developed a coaxial helicopter—that is, a helicopter with two sets of rotor blades on the same mast; the two sets turn in opposite directions. Coaxial rotors, while complex, eliminate the torque effect and so eliminate the need for a tail rotor. Pescara invented a mechanism to control cyclic pitch (tilting the rotor's plane of rotation fore and aft or side to side), so his helicopter could move forward as well as to either side and backward. That Pescara flew his cyclic-pitch-equipped, coaxial-rotored helicopter is all the more remarkable given that he used wire-braced biplane rotor blades, each set of which looked very similar to the Wrights' original glider. Pescara used four of these double blades per rotor and had two rotors whirling atop his craft. In the early 1920s, he was able to fly for more than a kilometer and hover for ten minutes at a time. But even though it was an important milestone, Pescara's craft was not practical because control was marginal and he could only fly forward at a pace equivalent to a brisk walk.[25]

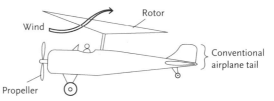

Figure 8.4 Autogiros have passive rotors tilted back slightly. The wind passing through them from below keeps them spinning.

One problem that Pescara attempted to solve is what happens when a helicopter's engine dies. If an airplane engine fails, the airplane can glide. Is there anything comparable that a helicopter can do? Pescara pointed out that the rush of air through a descending helicopter's rotor could keep the rotor spinning if the rotor were free to freewheel when the engine dies (and if the blades' angle of attack or pitch could be sharply reduced). Then, when the helicopter gets close to the ground, if the pilot quickly increases the rotor's pitch, the inertia of the spinning rotor might be enough to allow a brief increase in lift to brake the descent and produce a survivable, if not soft, landing. This trick is called autorotation, and today it is a standard emergency technique for helicopters. Pescara incorporated a primitive form of collective pitch on his helicopters and was probably the first to attempt autorotation.[26]

A different kind of flying machine, the autogiro, played an important role in helicopter evolution in the late 1920s and 1930s. An autogiro flies with an unpowered rotor. The rotor works something like a windmill: an autogiro has a separate propeller engine to push it forward; and by tilting the rotor back a bit, the air flow past the rotor keeps it spinning (fig. 8.4). In fact, an autogiro is always doing what a helicopter does when it autorotates.[27]

Autogiros take off like conventional airplanes but can land almost vertically; and using autogiros, pioneers such as Spaniard Juan de la Cierva figured out techniques to control and stabilize rotors that designers soon applied to helicopters. At first, autogiros had stubby wings with ailerons for roll control, but later ones had sophisticated cyclic and collective pitch controls as well as hinged, or flapping, rotor blades.[28]

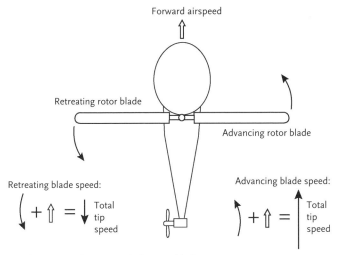

Figure 8.5 Advancing and retreating helicopter blades (top view, rotor turning counterclockwise). The helicopter's forward flying speed adds to the rotational speed of the advancing blade and is subtracted from the speed of the retreating blade.

When prototypes of autogiros (or helicopters) with rigid or unhinged blades fly forward, they have an alarming tendency to roll to one side. When a blade on one side moves forward, or advances, the craft's forward speed is added to the rotor's turning speed, thus increasing the blade's airspeed (fig. 8.5). When a blade swings around to the other side, it moves backward, or retreats, so the craft's forward speed subtracts from the airflow over the retreating blade. Advancing blades, with faster airspeeds, produce more lift than retreating blades do. So over the whole rotor disk, more lift is produced on the advancing side and less on the retreating side, which causes the rolling tendency. Designers such as de la Cierva discovered that if they hinged their rotor blades so that advancing blades could flap up and retreating blades could flap down, lift was equalized on both sides, and the rolling tendency disappeared. These discoveries were also directly applicable to helicopters.[29]

But even with control methods borrowed from autogiros, helicopters still had stability problems. In the mid-1930s, European inventors built helicopters with two opposite-turning rotors to overcome torque, with improved control but limited stability. Louis Breguet's company built a coaxial helicopter prototype with impressive performance (seventy-five miles per hour as a forward speed and flights of more than an hour), but it was destroyed due to control problems in autorotation. Just before the start of World War II, Heinrich Focke built a side-by-side twin-rotor helicopter with even more impressive performance.[30] European wartime priorities shifted away from helicopter research, however, and later progress in helicopter development shifted to the United States.

The Europeans had shown that true VTOL helicopters were possible, but Americans had to figure out how to make them practical and put them into production. Several companies began helicopter research programs just before the United States entered World War II. After trying almost a dozen different configurations, Igor Sikorsky's group settled on a single main rotor with a small antitorque tail rotor; and his researchers eliminated many of the bugs in pitch-control and rotor-hinge design.[31] In the 1920s and 1930s, Arthur Young, working on his own, had built several flying helicopter models to develop a stabilizing system. After joining Bell Aircraft, he applied his stabilizer to full-size prototypes. His stabilizer was a bar with weights on each end mounted on the rotor head at right angles to the two-bladed main rotor. It acted as a gyroscope to reduce unintended tilting of the main rotor.[32] Young's stabilizer led directly to the first commercially licensed helicopter, the bubble-cockpit Bell Model 47 familiar to viewers of the *M*A*S*H* movie and television series. Stanley Hiller started out by developing coaxial helicopters but concluded that the single main rotor and tail rotor configuration was more practical.[33] By taking Young's stabilizer bar a step further, he developed a breakthrough that led to truly stable helicopters. Rather than weights, Hiller put paddles on the bar at right angles to the main rotor. These paddles look like short pieces of rotor blade stuck onto the end of the cross bar. The pilot's cyclic and collective controls are linked to the paddles rather than the main rotor so that the pilot flies the paddles and the paddles, in turn, adjust the main rotor. This innovation greatly improved stability and significantly reduced the amount of effort a pilot needed to exert in order to move the controls. Hiller's prototype was so stable that his company published a picture of the craft hovering on its own, with sandbags in the seat and the pilot standing alongside with his hands in his pockets![34]

Other Americans continued with designs using two main rotors. Charles Kamman produced a line of helicopters with side-by-side rotors that had the rotor shafts mounted very close together, with each tilted slightly to the side so that the blades intermeshed. He called these compact, reliable choppers "synchropters."[35] Frank Piaseki pioneered large helicopters with tandem (twin fore-and-aft) rotors that led to Boeing Vertol's line of tandem-rotor heavy cargo helicopters used today by the U.S. Army and the Marines.[36]

In the end, no single person or group that can be identified as having invented the helicopter. Many designers contributed to the development of practical helicopters because researchers had to overcome several succeeding hurdles: first, power limits; then control problems; finally, stability problems. We now have practical helicopters in several different configurations, but are they the best approach to hovering and vertical flight?

The Down Side of Helicopters

Unlike flapping animal wings, the fixed wings of conventional airplanes are useless for hovering. A helicopter rotor can be effective for hovering, but it faces inherent problems when it moves forward. Because a practical helicopter needs to fly forward to be useful, are helicopters the most practical craft for VTOL? What are the pros and cons of using rotors, and are there alternatives?

We have already encountered the torque problem: any chopper with a single engine-driven rotor tends to rotate the fuselage in a direction opposite from the rotor's rotation. Coaxial rotors eliminate torque because the torques from the counter-rotating rotors cancel each other, but at a significant increase in complexity and weight. An increase in complexity is no small thing, given that helicopters are already the most complex and maintenance-hungry of flying machines.

Torque can also be avoided if the blades themselves provide their own thrust. Over the years, designers have come up with a number of schemes to put propellers on rotor blades; but none have worked very well. Others have tried routing jet engine exhaust through hollow rotor blades to tip outlets or mounting tiny jet engines on the tips of rotor blades, but the problems of sending jet exhaust or fuel from a nonrotating mast to a rotating rotor blade has never been adequately solved. Even though the tail rotor eats up about 15 percent of the engine's power in a single-rotor helicopter, designers have decided that this is an acceptable trade-off against weight and rotor-head and rotor-control complexity.

Helicopters face an inherent speed limit, even more restricting than the inherent limit on propeller-driven airplanes (see Chapter 3). As we have seen, when a helicopter flies forward, the craft's speed is added to that of an advancing rotor blade and subtracted from that of a retreating blade (fig. 8.5). If a helicopter flies too fast, speed causes problems on both sides. First, on the retreating side, the blade's airspeed will drop so low that it will stall, eliminating much of the lift on that side of the rotor disk. Second, the airspeed of the blade on the advancing side will become so high that the outer part of the blade will go supersonic, reducing lift and vastly increasing drag. The resulting drag on the advancing blade will place an unbearable power burden on the engine. These two effects create a catch-22 for the rotor. If a helicopter could somehow manage to keep flying with rotor blades stalling on the retreating side and going supersonic on the advancing side, the resulting massive vibrations would quickly destroy the rotor hub, if not the whole craft.

Even with very powerful turboshaft engines (the helicopter equivalent of turboprop engines), very few helicopters cruise faster than about 160 miles per hour. Although a few specially prepared helicopters have reached almost 250 miles per hour, 200 miles per hour seems to be the practical upper limit of production helicopter cruise speeds.[37] In contrast, turboprop cargo airplanes (for example, the Lockheed C-130 Hercules) or passenger airplanes (such as the Lockheed Electra or the Beechcraft 1900) typically cruise almost twice as fast, in the range of 300 to 400 miles per hour.[38] So the VTOL benefit of flying with rotors comes with a huge penalty in speed.

Other Ways to Hover

At about the time that practical helicopters were being perfected and put into production, inventors began to suggest ways to get around the helicopter speed limitation. The most successful is the tiltrotor aircraft, sometimes called a "convertiplane." The idea with the tiltrotor is that the craft takes off vertically with rotors mounted on the ends of airplane-like wings. Then the pilot tilts the rotors forward so they become propellers, and the craft flies like an airplane. While hovering or flying vertically, it operates as a side-by-side twin-rotor helicopter. It starts flying forward like any helicopter (by tilting the rotor blades forward a bit); but as it accelerates, its wings start to produce lift. As the wings produce more and more lift, the rotors can tilt down farther and farther until they are completely horizontal and producing only thrust, operating entirely as propellers. In the propeller configuration, the rotors no longer have advancing and retreating blades, so the craft is subject to the much higher propeller speed limits rather than the lower forward speed limits of a helicopter rotor.

Although this conversion may sound simple, the devil is in the details. Controlling such a craft in hover or vertical flight is quite different from controlling it in forward flight, and the transition from one mode to the other can be even more challenging. Developing a control system that works in both modes, as well as during the transition, is a major design challenge. The combination of the tilt mechanism and the dual-mode control system is even more complex than that of a conventional helicopter.

As early as 1955, Bell Aircraft produced a small, experimental tiltrotor machine, the XV-3, powered by a single piston engine driving two rotors through a gearbox and long shafts to the wingtips. The XV-3 successfully hovered and transitioned to forward flight with fully tilted rotors.[39] Other experimental tiltrotors followed, the most recent of which was Bell's highly successful XV-15, with two turboprop engines, each mounted in a tilting pod on a wingtip. The entire engine and rotor assembly tilts up to fly as a helicopter or forward to fly as an airplane. The XV-15 was reputedly relatively easy to fly, so

Bell's team actually invited a number of guest fixed-wing and helicopter pilots to fly it. After less than an hour of ground instruction, the guest pilots all were able to hover the XV-15 and transition it to forward flight.[40] Based on this success, Bell proposed scaling up the XV-15 to produce a vehicle to replace military medium-lift helicopters. The result is the V-22 Osprey, which is poised to replace medium-lift helicopters in the U.S. Marines and the Air Force.[41] (The U.S. Army also considered buying V-22s but backed out due to the high cost and is instead buying an updated version of the venerable, Vietnam-era CH-47 Chinook.) The Osprey has a longer range, larger payload, and about twice the speed of the thirty-year-old helicopters it will replace. It is, however, very expensive—nearly twice the price of a similarly sized helicopter—and very complex, and it faces unresolved questions about reliability and maintainability.[42]

One final method of hovering, a sort of brute-force approach, is worth considering. This is hovering by direct engine thrust. In principle, if an engine is powerful enough, a pilot ought to be able to point the engine straight up and use it to hover. In fact, some extremely powerful, piston-engine aerobatic airplanes such as the Sukhoi Su-26 and the Extra 300 can hover this way. Such "prop-hanging" is essentially an air-show stunt with no practical use because the airplanes certainly cannot land by sitting down on their tails. (Experimental aircraft of the 1950s, such as the Convair XFY-1 Pogo and the Ryan X-13 Verti-Jet, were designed to take off and land this way but were extremely challenging to land and never became practical.) As further proof that small size is an advantage in hovering, a large segment of radio-control models routinely hover on the propeller. This prop-hanging ability has even spawned a new category of maneuvers, "3-D aerobatics," and models capable of performing these maneuvers appear to be the fastest-growing branch of the hobby today. Even these overpowered models, however, cannot land or take off this way.

So far, the only production airplane routinely capable of hovering and landing vertically using engine thrust is the Harrier jump-jet. The original Harrier was a British design built by Hawker-Siddeley, and the current updated Harrier II (known to the U.S. Marines as the AV-8B) was modified and improved by McDonnell-Douglas in the United States. The Harrier is a relatively conventional-looking, single-seat, single-engine light jet attack plane (a small bomber). Its engine, however, is far from conventional. Its massive Pegasus turbofan engine blows compressor bypass air and jet exhaust out through pairs of rotating nozzles, a pair in front for compressor air and a pair in back for jet exhaust. These nozzles give the pilot what engineers call "vectored thrust." The pilot can rotate the nozzles from straight back to straight down. When the nozzles point straight back, the airplane operates as a perfectly conventional jet airplane. When the nozzles point down, however, they produce enough thrust so that the craft can hover and take off or land vertically. Even the mighty Pegasus does not have enough thrust to push up the Harrier vertically with a full load of fuel and munitions; but if a pilot tilts the nozzles partly down and partly back, the fully loaded Harrier's take-off run is only about half as long as it would be with nozzles straight back like a conventional jet's. At the end of a flight, having dropped its bombs and used up much of its fuel, the Harrier can land either conventionally on a runway or vertically by hovering on vectored thrust.[43]

When a Harrier hovers, its conventional control surfaces don't work, so it uses a set of small jet nozzles on the wingtips and tail. These nozzles use bleed air from the engine's compressor to maneuver the craft while hovering, functioning in a way similar to the reaction jets of a spacecraft in outer space. Having seen Harriers perform several times at air shows (definitely worth the price of admission), I have witnessed their helicopter-like hovering ability: they can fly sideways and backward and pirouette in place. One thing they are not is stealthy—the noise of a hovering Harrier is painful up close and can probably be heard for miles in open country.

The Harrier's basic design is more than four decades old. It combines the best features of helicopters (vertical take-off and landing) with jet airplane speeds—subsonic but only barely, at more than six hundred miles per hour. The U.S. Marines have used it for more than two decades, the British for even longer. A number of vectored-thrust, jet VTOL craft have been designed since the Harrier debuted, but apparently only the Soviet Union's Yakolev Yak-38 went into production, and it was not a success. (Intended to operate from aircraft carriers, the Yak-38 was plagued by a miniscule payload, absurdly limited range and duration, highly unreliable engines, and the apparent inability to carry any munitions at all in hot weather.)

If the Harrier was successful, why haven't more vectored-thrust jet aircraft been developed? One reason may be the sheer power required. The engine must be able to produce thrust greater than the airplane's weight. Although a few of today's frontline jet fighters have engines that produce more thrust than their own weight, the vast majority of airplanes fly on much less power. Most fly with engines producing thrust equivalent to much less than half their weight. More powerful engines are both more expensive and heavier and only justified for extremely high-performance combat aircraft. Even fighter jets with high thrust-to-weight ratios cannot land vertically because they have no way to sit down on their tails and no way to maneuver while backing down slowly.

Other drawbacks to jet-powered VTOL are complexity and fuel requirements. Hovering on vectored thrust requires a separate set of control mechanisms from the ones used for conventional flight, which adds a whole new system. Furthermore, the hovering control system should ideally use the same cockpit controls as the conventional flight controls to keep the pilot workload manageable. Combining these two control systems is a major integration challenge. Finally, hovering on vectored thrust is far and away the most inefficient form of hovering. For example, the fuel requirements of hovering and vertical flight are massive and greatly reduce the Harrier's range for a given fuel load, so pilots usually use vectored thrust to shorten conventional take-off and landing rolls and reserve vertical landing for situations in which off-airport operations are vital. This inefficiency means that civilian use of vectored jet thrust for VTOL craft is unlikely because no civilian missions exist that would justify the huge purchase price and fuel costs.

Combat, however, may well provide ample justification. The British used the Harrier with notable success in the Falkland Islands war with Argentina, where it surprised many experts as an adequate air defense fighter in addition to its expected role as a bomber. The

U.S. Marines are also fond of the off-airport capabilities of their Harriers, so I am somewhat perplexed that no other successful military vectored thrust jets have come on the scene during the Harrier's long lifespan. The U.S. military is currently testing a new attack jet, the F-35, one version of which will have VTOL capability, but the jet will not be operational for several years to come.

Hummingbirds: More Harrier Than Helicopter

Oddly enough, when animals use conventional hovering, they are doing something not terribly different from the prop-hanging of aerobatic airplanes. At its simplest, conventional hovering is almost like pointing the body straight up and attempting to fly up instead of forward. A key difference between a hovering hummingbird and a prop-hanging Extra 300 (or a hovering Harrier, for that matter) is that the hummingbird is still actively using its wings, whereas the wings of an Extra 300 (or a Harrier) are so much dead weight in a hover. The flexibility, both literally and figuratively, of animal wings give animals a huge advantage in the vertical flight arena.

The key to the animal system is versatility. Flapping flight lends itself to variation in wing movements and wing shape so that the same wings can function reasonably well as helicopter rotors (hovering), combined wings and propellers (forward flight), and simple wings (gliding). This versatility is essential for flying animals. Most need to be able to land and take off in tight quarters or from perches and fly in cluttered, obstacle-strewn environments. The flapping propulsion mechanism may not use energy most efficiently, but it gives birds or bats or bees unmatched ability to fly with great agility over a huge range of speeds, down to and including hovering.

In contrast, flying machines that hover—so they can land and take off vertically—are inherently highly specialized. The simplicity of the fixed-wing airplane design precludes hovering in all but the rarest of extreme cases. Helicopters are the only really practical machines for hovering and vertical landing, using the aerodynamic lift of rotors more efficiently than the thrust from jets or propellers. Even helicopters, however, are quite specialized, so we use them only in situations in which their hovering ability is a great advantage—for example, in rescues or for landing on offshore oil rigs. They are simply too complex, too expensive to operate, and too slow for us to use them for mundane tasks such as carrying cargo or passengers.

In human aeronautics, specialized tasks call for specialized machines, but general tasks call for simple machines. Flying animals, in contrast, cannot switch from one machine to another to fly in different ways. Each individual has one basic design, and the animal needs to be able to carry out all its important tasks using its one and only body. Nevertheless, this basic design is not "fixed" in the same way that we would call an airplane's or a helicopter's design "fixed." The animal can use its flexible, adjustable body parts—including wings—to reconfigure itself to perform different tasks: from forward flight, to hovering, to walking on the ground. Flying animals may be limited in speed and efficiency, but no flying machine approaches a bird's versatility.

~ 9 ~

Stoop of the Falcon
Predation and Aerial Combat

Peregrine falcons are one of the few birds that attack other birds in flight. Falcons are also one of a small number of animals that thrive in highly urban areas. Never common to begin with, after World War II their numbers dropped even more dramatically because of reproductive problems related to the widespread use of the insecticide DDT. After DDT was banned in the early 1970s, conservation organizations set up captive rearing programs for falcons (and other birds affected by DDT) and soon began releasing falcons back into nature. Strangely, "back into nature" for falcons includes the centers of major cities.[1] What makes downtown Baltimore or Minneapolis so attractive to these birds?

Peregrine falcons obtain food by attacking other birds in flight. They stoop, or dive steeply, onto their prey, dropping from as much as several hundred feet above their intended target. In this way, the selected victim is less likely to see the falcon, and the falcon can build up tremendous speed in its dive. The falcon's feet slam into its prey in midair, striking hard enough to stun or kill the other bird, which the falcon then carries off to a perch or its nest to eat.

At least two things about big cities attract peregrine falcons. First, they normally nest on cliff ledges and also seem to like skyscrapers, where their nests are protected from other predators and the falcons have a fabulous view of potential prey. Second, pigeons are a typical prey for falcons, and most big cities in North America and Europe are overrun (overflown?) with pigeons. Programs that release captive-reared falcons in big cities have been quite successful, and a few falcons have even decided to move into cities on their own. Falcons seem to thrive in the urban environment, much to the pleasure of birdwatchers and people who work or live in the buildings where falcons nest.[2]

The peregrine falcon is a classic aerial predator: it cannot use concealment to sneak up on its prey, as a tiger can in the underbrush. Ideally, an aerial predator should have excellent vision (as good as or better than its prey) and be faster, more agile, and stronger than its prey. The task of intercepting another moving object in three dimensions is mentally challenging, so the predator needs to be either very smart or in possession of very sophisticated chasing reflexes. Because aerial predation is such a challenging way to catch food, these predators are rather rare in the animal kingdom, at least relative to the abundance of potential prey.

Aerial predation has a close parallel in aviation. Air-to-air combat requires almost the same skills and involves a very similar task: aerial interception, with a potentially fatal outcome for the target. And in both nature and aviation, an arms race often develops between the attacker (the predator or the fighter airplane) and the victim (the prey or the bomber). If the attacker starts to gain the upper hand over the victim, the victim's survival depends on developing a counter-ability to defeat the attack. If falcons were to become a serious threat to, say, mourning doves, the doves might evolve better vision or more erratic flight to counter the threat. If doves were an important prey for falcons, then the falcons in turn would have to develop new flight abilities or hunting tactics. The same arms race occurs in combat aviation, when increases in speed, maneuverability, or altitude of the victim lead to compensating changes in the attacking airplanes, and vice versa.

Aerial Predation

A predator is any animal that catches and eats prey animals, and a huge number of species of birds, bats, and insects are predators, although relatively few catch prey on the wing. A few predators attack prey as big as themselves, but most are a good bit larger than their prey. Familiar examples of predators include owls, hawks, dragonflies, tiger beetles, ladybird beetles (ladybugs), and praying mantises. Some predators—for instance, insect-eating woodpeckers, warblers, and most bats—eat animals a fraction of their size.

All of the predators I have mentioned are perfectly capable flyers; and some, such as owls, do attack terrestrial prey from the air. Most, however, land and chase their prey (as ladybird beetles and warblers do) or ambush their prey on the surface (as praying mantises do). Why is aerial predation fairly uncommon even among predators that can fly? The biggest challenge to hawking, or attacking prey on the wing, is that a flying predator has nowhere to hide. A cougar can sneak through underbrush to get close to a deer before pouncing, but a falcon doesn't have that option. In principle, a vigilant prey animal that is flying in the open can look in any direction and see a faraway predator. Ducks, for instance, use eyes set on the sides of their heads to watch for foxes and bobcats on the ground as well as falcons in the air. Because their field of view is more to each side than to the front, they are well adapted to watch for predators approaching from any angle.

Because aerial predators cannot get close to prey by sneaking through cover, they must be able to fly much faster than their prey; and one way is to be substantially bigger. Thanks to the way in which flight speeds scale with size (see Chapter 2), bigger flying animals usually fly faster than smaller ones do. Dragonflies, although quite fast for their size, still tend to prey on much smaller insects, such as mosquitoes and midges. Dragonflies can and do occasionally chase down insects almost as big as themselves, but usually they go after smaller prey. If the sizes are different enough—say, a nighthawk chasing a small moth—the attacker can have enough of a speed advantage to gain some of the benefits of an ambush. The nighthawk can close with the moth so fast that even if the moth sees the bird some distance away, it cannot fly fast enough to run away.

Another trick to gain a speed advantage is to dive steeply at the prey, using the falcon's stoop technique. The multiple advantages of this method include starting so far above the prey as not to seem threatening or even noticeable. Also, in a steep dive, gravity can do much of the work of acceleration, so that a stooping attacker can reach speed several times faster than in level flight. Finally, the speed of the dive gives the predator a somewhat ambush-like attack, so the prey has little time to notice and evade.

Another challenge for aerial predators is that flying animals tend to be extremely maneuverable. Over many millions of years—at least 150 million for birds and more than 300 million for insects—flying animals have evolved great agility, largely to avoid both terrestrial and aerial predators. When it comes to maneuvering, prey actually have the upper hand: all else being equal, a slower, lighter flyer can make tighter turns than a faster, heavier flyer. Of course, in real life, "all else" is never equal, but the challenge remains. Aerial predators must be significantly faster than their prey to have any hope of interception. Because prey animals are likely to be agile, a predator must be not only faster but much more maneuverable for its size than the prey; and even then, it may not be quite as maneuverable as smaller, slower prey animals.

Consider a peregrine falcon stooping at 120 miles per hour on a small duck flying at 40 miles per hour. If the duck turns sharply at that speed, the falcon needs to use three times the duck's turning effort and will then experience three times the aerodynamic load on its wings and body. In fact, if the duck turns as hard as it can, the falcon will not be able to turn sharply enough to literally follow the duck through its turn. In many cases, however, the falcon does not need to. If the duck unwisely chooses to turn as soon as it is under attack, at a point when the falcon is still flying well above it, the falcon simply needs to adjust its direction enough to intercept the duck on its new heading rather than actually follow the duck through its turn. In other words, the falcon can cut the corner. The duck is in a situation similar to that of a fighter pilot trying to evade a much faster guided missile. If the duck (or the fighter pilot) turns too soon, the falcon (or the missile) will have enough time to correct for the course change by cutting the corner. The trick is to wait until the last possible second to turn. Then, when the would-be victim finally turns, the attacker is too close to adjust for the target's new heading and too fast to follow the target through its turn.

Biologists see the relationship between predator and prey as a classic case of coevolution. The prey evolves greater maneuverability and ever more precise escape behaviors, forcing the predator to become both more maneuverable and more intelligent so as to be able to predict and react to the prey's responses. Demonstrating the similarity between the analogous changes in nature and combat aviation, evolutionary biologists use the term *arms race* as shorthand for the coevolution of a predator and its prey.

Rogues Gallery of Aerial Predators

Peregrine falcons are probably the best-known members of the falcon family. Some members of that family prey on mice, lizards, and other terrestrial prey; but others, including the large gyrfalcons, use the peregrine falcon's stooping attack to prey on other birds. Indeed, falcons have given their name to *falconry,* the ancient technique of taming, training, and hunting with birds of prey. Although other hawks and even eagles have been used in falconry, the falcon's immensely fast, surgically precise diving attack on game birds is the archetype.[3]

Although falcons are probably the most specialized aerial predator of other birds, hawks, owls, and eagles may attack flying birds as well. Such attacks are usually opportunistic, a matter of being in the right place at the right time rather than a specialized hunting tactic. The frigatebird also attacks other birds in flight, but not to eat the victims. This seabird is an oddity: it doesn't swim (its feathers are not waterproof), so it cannot dive into the water to fish, although it can pluck small fish from the water as it skims over the surface. The frigatebird attacks smaller seabirds such as gulls and boobies as they return to their nests with crops full of fish, harassing its victim until the smaller bird regurgitates its fish, which the frigatebird then catches and eats in midair.[4]

Little Night Monster

Although quite a few birds fly at night, they are generally immune to aerial predation because avian predators are highly visual. In the dark, a falcon has no hope of maneuvering precisely enough to catch another bird. Until recently, in fact, biologists had not found any aerial predator of nocturnally flying birds. Now, however, a truly amazing tale of aerial predation has come to light. Biologists think that a rare Mediterranean bat actually preys on nocturnally migrating songbirds, making it the only known predator of night-flying birds.

Many small migrating birds fly long distances at night, even though most are diurnal when not migrating. The main reason they migrate at night is presumably to avoid daytime predators such as falcons. A few years ago, Spanish and Italian biologists discovered feathers in the droppings of the greater noctule bat (also called the giant noctule). Carlos Ibáñez and his Spanish colleagues learned that the bat droppings only contained feathers during the times of heavy bird migrations. They concluded that the greater noctule preys on migrating songbirds during the spring and fall and on insects during the summer. (Like many temperate bats, the greater noctule probably hibernates during the winter.) This conclusion seemed so unlikely that other skeptical biologists suggested that the bats were just accidentally eating shed feathers, mistaking them for insects. (Why a bat would then go ahead and swallow an accidentally caught feather was left unexplained.) Ibáñez's group proceeded to use blood chemistry to show convincingly that greater noctules eat birds during the fall and spring, when the birds are migrating, and only insects in the summer.[5]

Why not just watch the bats to settle this question? First, the greater noctules hunt only at night, so direct observation is difficult. Moreover, migratory birds fly several thousand feet above the ground, so the chance of seeing such an attack is remote even in daylight, let alone at night. Finally, greater noctules are among the rarest bats in Europe. Directly observing their attacks is simply not practical.

Biologists know something about the predatory attacks of bats closely related to the greater noctule. They typically hunt at night in open air well above the forest canopy, using echolocation, or sonar, to find and intercept prey (see Chapter 6). Rather than returning to the perch to eat prey, they always eat on the wing; and greater noctules probably eat birds in the same way. These bats are the largest in Europe, with a body weight about three-fourths that of a robin's but with a slightly larger wingspan. They also have impressively large teeth for their size and, at fifty grams, weigh more than twice as much as the songbirds that typically migrate through their haunts. According to a radio interview with Ana Popa-Lisseanu, one of Ibáñez's colleagues, greater noctules probably hunt birds by flying high above their migration routes and locating their prey with echolocation. Once the bat locates a victim, it dives on the bird and traps the animal in its wings. While falling, the bat quickly bites away the high-value parts of the bird (mainly the fat- and protein-rich breast muscle) and then drops the rest of the carcass and flies off before getting too low.[6] (Because the bats are so rare and scavengers so common, trying to find these dropped carcasses would be futile.)

Greater noctules live along the northern coast of the Mediterranean Sea, including parts of Spain and Italy, where migratory birds are funneled to narrow crossing points across the sea. Birds cross these areas by the billions and are a rich resource for potential predators. Full of fat to fuel their journey, vast numbers fly along predictable routes at predictable times of the year. As far as we know, however, the greater noctules are the only predators that have evolved behaviors to take advantage of this resource and are also the only bats known to attack birds in flight. Fortunately for the birds, greater noctules are so rare that their effects on migratory populations are negligible.

Batting at Bugs

Insects are more typical fare for bats as well as for the majority of birds that are aerial predators. These insect eaters are obviously much bigger than their prey, an extreme example of the tendency of animal attackers to be larger than their victims. In contrast, fighter airplanes tend to be smaller than (or, at most, equal in size to) their targets in combat aviation.

Some bats are gleaners, plucking insects from foliage as the bat hovers or flies past. Most bats hawk insects (catch and eat them in flight) at least some of the time, and many use hawking as their main hunting tactic.[7] The most specialized hawking bats have long, narrow wings for fast, efficient flight in open air well above cluttered environments such as forests. Others have shorter wings for slower, more maneuverable flight and hawk insects within forests. The high flyers tend to have longer-range sonar and to make a single capture attempt each time they pass a target. Those hawking in cluttered environments tend to have shorter-range but more sensitive sonar and often make several capture attempts as they zig-zag through a swarm of insects. The high flyers, being fast, can cover more distance in a search, whereas the forest flyers, being more maneuverable, are better at following the twists and turns of evading prey.[8]

Biologists long assumed that insect-hawking bats simply snapped up their prey by mouth. But in recent years, researchers have employed sophisticated photography and night-vision equipment to document other ways in which bats collect insects. Many use their wings like tennis racquets, deftly tapping an insect to deflect it into their mouths. One European bat, Daubenton's bat, flies low over water or open meadows and collects insects with the tail membrane stretched between its hind legs. The bat swings its tail membrane down and forward to cup insects as it flies and then ducks its head to grab an insect in its mouth. Other species use a wing to knock the insect into the cupped tail membrane. As it ducks its head to pluck an insect from its tail membrane, a bat often does an aerial somersault as it gulps down the morsel.[9]

Bats Aren't the Only Ones

Thanks to their sonar, bats have monopolized the nocturnal insect-hawking niche, but during daylight hours a diverse group of birds takes over. Swifts, swallows, and nighthawks are specialized insect hawkers, and many other birds occasionally take insects in flight or dart from a perch to nab them. Swifts and swallows, although not as closely related as once thought, have evolved similar body forms for similar lifestyles—an example of parallel evolution, in which similar life histories have led to selection for similar bodies. Both have short, streamlined bodies and long, pointed wings for fast, efficient flight.

The vast majority are insect hawkers, although some tree swallows will eat berries when bad weather grounds insects.

People often describe the flight of swifts and swallows as batlike (although, because the birds have been around longer, we probably ought to call bat flight swift-like). Most swallows have long, deeply forked tails, as opposed to the swifts' stubby ones; and swallows often intersperse brief glides between bouts of flapping.[10] Both swifts and swallows are so thoroughly aerial that they drink and even bathe in flight. They swoop down over the surface of a pond or a river and dip their beaks to drink or drag their bellies to bathe. Swifts even copulate on the wing, and biologists have some evidence that they may actually sleep in flight.[11]

Swifts and other birds that hawk insects use more oral and less athletic food-handling methods than bats do. They tend to have short, wide beaks with enormous gapes—that is, the ability to open their mouths extremely wide. As far as we know, they collect prey only by mouth, without any wing batting or tail cupping. (Feathered wings and tails don't lend themselves to such manipulations.) Any of these birds can forage at any time and at any altitude, but they tend to divide up the environment in both time and space. Swifts hunt most intensively in the middle of the day, swallows focus on midmorning and midafternoon, and nighthawks hunt at dawn and dusk (not, however, overnight). Any of them can be found inches over a pond or thousands of feet in the air, but swifts tend to hunt at higher altitudes and swallows a bit lower. Nighthawks often start high and gradually descend as they forage.[12] In spite of these general tendencies, all of these birds (and bats as well) take advantage of places where insects collect, such as in the lee of fences, treelines, buildings, and cliffs.[13]

Insect Falconry

Scientists have named more species of insects than all other animals combined, so amid that diversity, we ought to find some aerial hunters. Many flying insects are predators, but only a few take prey on the wing. The most highly specialized of these are undoubtedly dragonflies and their close relatives, damselflies.

Dragonflies are sometimes called living fossils because they evolved their basic body form more than 200 million years ago and have not changed much since. A dragonfly has two pairs of large, high-aspect-ratio wings, giving it fast, efficient flight and making large dragonflies among the fastest of flying insects. Moreover, because they can adjust the wing beat of the front and hindwings almost independently, dragonflies are among the most maneuverable of insects. Their legs are so specialized for grabbing and caging prey that they can barely walk: to shift its position as little as an inch or so, a dragonfly will fly rather than walk.

Dragonflies use their speed and maneuverability advantages to chase down and catch smaller insects. Some of their prey insects are themselves quite agile, such as flies, mosquitoes, midges, and wasps.[14] The fact that dragonflies are common today is a testament to the effectiveness and adaptability of their ancient design, even though their modern prey is quite different from their ancestors'.

Surprisingly few other insects follow a hawking lifestyle. Robber flies, or asilids, distant relatives of horse flies, are fairly common and feed by catching prey on the wing.[15] They are similar to a horse fly in size but more slender, and I have always found their appearance to be menacing. Large robber flies will sometimes attack prey as large as or larger than themselves, and they eat well-defended insects such as bees, wasps, and dragonflies as well as elusive ones such as other flies.

A few kinds of insects attack other flying insects but, in the manner of frigatebirds, not with the intention of eating them. Adult conopid flies, or thick-headed flies, are not carnivorous; most are probably nectar feeders. Their larvae, however, are parasites of other insects such as bees and wasps. (Technically, they are parasitoids because they normally kill their hosts, which parasites rarely do.) The adult conopid catches its preferred host in flight and injects its eggs into the unfortunate victim. The eggs hatch, and the larvae grow and develop inside the host.[16]

The human bot fly is even stranger. Bot fly larvae are parasites of vertebrates, and human bot fly larvae develop just under the skin of humans. Although the adult bot fly is about the size of a horsefly, female bot flies have no way to get their eggs through human skin. So a gravid female bot fly chases and grabs a mosquito in flight. She glues her eggs onto the outside of the mosquito and then releases it. When the mosquito alights on a person to suck blood, the person's body heat causes the bot fly eggs to hatch. The bot fly maggot drops onto the person's skin and, after the mosquito leaves, burrows into the person's skin through the mosquito's feeding incision.

Incredible as this story may seem, it has even more complications. First, only female mosquitoes suck blood. Second, some species of mosquitoes are likely to bite humans, whereas others prefer nonhuman mammals or birds. Because bot fly larvae will die in the wrong host, does the mama bot fly know to choose not only certain species of mosquito but also only the females of those species? Or does she spread her eggs among many different mosquitoes and hope that one will hit the jackpot? So far, biologists don't have the answers to these questions. By the way, human bot flies are fairly common in the New World tropics but do not carry any diseases and so are disgusting nuisances rather than a significant health problem.[17]

Certain insects carry out a behavior similar to predation: aerial forced copulation. A few insects mate in flight; and among dragonflies, both partners must be willing. In other groups, however, males chase females and attempt to copulate regardless of the female's willingness or lack thereof. In honeybees, for example, a virgin queen leaves the hive on her nuptial flight and produces pheromones that attract drones over large distances. If other hives are in the vicinity, the queen soon finds herself at the center of a comet-like swarm of drones, all trying desperately to mate with her. She flies fast and evasively so that only the fastest and strongest drones get close enough to copulate. The queen mates with several drones on each nuptial flight and may make as many as five such flights over two or three days. She then never mates again but stores their sperm for use over the rest of her reproductive life.[18]

Some species of syrphid flies, or hover flies, use similar though more solitary tactics. Males hover near stationary objects—trees, shrubs, walls—and wait for females to pass by. They will immediately track and chase anything that appears to be the size and speed of a female hover fly—meaning that they spend a lot of time starting and abandoning chases of inappropriate insects. If the target turns out to be a female of the correct species, the male chases her down and, if he intercepts her, copulates, even without her cooperation.[19] Such "rape" matings are quite unusual among insects and may indicate that the male's interception task is very difficult. Males and females are usually similar enough in size to keep their arms race essentially tied; the male can never gain a significant edge. This equality may mean that successful interceptions of correct females are so rare that the male is desperate to copulate whenever he is actually successful.

Aerial Combat: Attacking Other Airplanes in Flight

A male hover fly's interception of a female fly is analogous to a fighter airplane's interception of another airplane, but the stakes of aerial combat are much higher. As in a predator-prey encounter, death is a likely outcome.

The idea of using airplanes to attack other airplanes was not obvious to military authorities before World War I.[20] At the beginning of that conflict, both sides used unarmed airplanes for what they called observing (what we would now call reconnaissance): flying over enemies to see what they are up to. When aviators from opposing sides encountered each other, they passed harmlessly, unable to do anything more damaging than shake fists at each other. Early observation pilots sometimes carried a few small bombs that they could toss over the side at particularly tempting enemy targets, but most were fairly ineffective.[21]

After a few months of such actions, both sides realized that they would rather not have enemy observers buzzing overhead with impunity. At first, aviators began carrying pistols and rifles so they could take potshots. Soon they mounted machine guns on observation planes and then moved on to mount them on airplanes specifically designed to attack other airplanes.[22] The airplanes that we now call fighters were developed within the first year of the war.[23]

By the middle of World War I, military airplanes had become specialized enough to fall into the general categories we still use today. Bombers were large airplanes to carry heavy bomb loads, usually with a crew of two or more. Observation (reconnaissance) planes were smaller than bombers but large enough to carry a pilot and an observer and perhaps a bulky camera or a radio set. Fighters were small, maneuverable airplanes designed to shoot down bombers or observation planes or to protect friendly bombers and observers by shooting down enemy fighters.[24]

Since World War I, through World War II's Battle of Britain and up to today, fighter pilots have been the military's glamour boys (and, more recently, girls)—the steely-eyed, highly trained elite among pilots. During World Wars I and II, they were often national celebrities. Fighter pilots are usually young, smart, fit, self-confident, and arrogant. The best students in each class of military pilot trainees—the cream of the crop—are chosen to fly fighters.

We tend to think of fighter pilots, like infantrymen, as the tip of the spear—that is, as frontline military offense. Yet in fact, when fighter airplanes engage in aerial combat, they are fundamentally defensive, not offensive, weapons. When a fighter is trying to shoot down other planes, it typically plays either an interceptor or an escort role. An interceptor tries to shoot down enemy bombers, thus defending friendly territory. An escort tries to shoot down enemy interceptors, thus defending friendly bombers. Bombers are actually an air force's offensive weapons; and bomber pilots, whom fighter pilots sometimes deride as "truck drivers," are the personnel who carry the fight to the enemy.

Jimmy Doolittle, who commanded most of the U.S. fighters based in England during World War II, championed a role for fighters that is arguably more offensive than defensive.[25] In a technique known as fighter sweeps, groups of fighters patrol enemy territory independently of bombers, seeking to attract and engage enemy fighters, catch them in vulnerable conditions (landing or taking off), or destroy them on the ground. Although the object is to achieve air superiority and destroy enemy airplanes, the ultimate purpose is still to protect the bombers. Doolittle claimed, in fact, that fighter sweeps saved more bombers than close escort tactics did, illustrating that, even when on the offense, a fighter's function is defensive. Moreover, fighter sweeps and other air-superiority tactics work best if a force has more and better fighters than its enemy and if the enemy is willing to engage in the absence of bombers.

Air combat is most similar to aerial predation when fighter airplanes act as interceptors—that is, when they attack bombers. Fighters are faster and more maneuverable than bombers are, just as aerial predators tend to be faster than their prey and, if not more maneuverable, at least highly maneuverable for their size. Yet even though both aerial predation and aerial combat involve arms races with increasing trends in speed and maneuverability, they also have some basic differences. Predators are usually bigger than their prey, partly because smaller prey are easier to subdue but also because, by being bigger, they are also faster. Being bigger is a hindrance for maneuvering, however. Therefore, unlike predators, fighters tend to be smaller than bombers so they can be more maneuverable. Rather than using large size as way to be faster (which doesn't work at jet speeds anyway), fighters use brute engine power to give them a speed edge over bombers. The cost of high power is high fuel consumption.

Fighters are notoriously "short-legged" (short-ranged) compared to bombers, which is why successful escort fighters are much less common than successful interceptors. Early models of Great Britain's World War II–era Spitfire and Germany's Me-109/Bf-109 were excellent interceptors but poor escorts: they did not have the range needed to escort contemporary bombers all the way to the target and back.[26] A major feature in the success of Japan's Zero and the U.S. P-51 Mustang was their extraordinarily long range (for fighters). Both were excellent in aerial combat for their day, but both also had much longer range than other contemporary fighters, making them outstanding escorts.[27]

When fighters attack other fighters, the analogy with predation breaks down further. Although two predators with similar abilities could theoretically attack each other, usually they shun such actions to avoid crippling injuries. When opposing fighters engage each other, they do so because pilots on both sides believe they have a chance of prevailing, which in turn means that their airplanes' flight capabilities and destructive power are approximately similar. Air combat between fighters is what fighter pilots train for and what top fighter pilots aspire to. Assuming that opponents are reasonably well matched, the situation requires extremes of skill, intelligence, and daring.

The closest biological analogy to fighter-on-fighter combat may be territorial defense actions among male dragonflies or male birds. The opponents are well matched in aerial ability, and the outcome is determined by whichever animal has a slight edge in skill, motivation, experience, or some combination of these qualities. The stakes are not quite as high as they are in predation or air combat: the vanquished male retreats with his life if not his dignity intact. These confrontations are more ritual than fight, and both parties are content to show off their abilities with minimal bodily contact rather than risk injury in an actual brawl.

Another significant difference between aerial combat and aerial predation is in detection range. To locate its prey, a predator relies on vision (or echolocation, in the case of bats). Fighter pilots in World Wars I and II also relied mainly on vision, although World War II pilots sometimes got help from surface-based radar and some night fighters were equipped with primitive airborne radar. Along with the development of jet fighters came great advances in airborne radar, and for the past three decades or so, all state-of-the-art fighters have carried radar.

Radar gives a fighter pilot the ability to detect other airplanes far outside his or her visual range, in some cases more than one hundred miles away. In principle, radar allows a fighter pilot to engage and attack enemy airplanes dozens of miles away—well beyond visual range—using radar-guided missiles. In practice, however, as pilots learned in the Vietnam War, such attacks almost never happen.[28] In spite of sophisticated devices to automatically identify friendly airplanes (known as "identification, friend or foe," or IFF, signals), pilots frequently need to visually identify potential targets to ensure they are enemies. Moreover, the first few generations of radar-guided missiles required the launching fighter to keep its radar locked on the target airplane. These were not terribly reliable and had a low probability of hits at long range.[29] Today's IFF systems and missile reliability are much better, and modern airborne radars let pilots locate and track other airplanes at amazing distances. Even so, the good old eyeball is still important for detecting and responding to threats, especially for distinguishing enemy from civilian or neutral airplanes.

Countermeasures: The Prey Fights Back

Sometimes the victims of an aerial predator or an intercepting fighter are relatively helpless, such as, for example, a falcon attacking a newly fledged robin just out of the nest or a fighter jet attacking an unarmed cargo plane. In other cases, however, the victim has some means of thwarting the attacker's assault.

As a rule, flying animals are extremely maneuverable (see Chapter 4), at least partly because of selection pressure to evade predators. Maneuverability is one element in Mother Nature's arms race: as predators evolve maneuverability, prey species are under selection pressure to become even more maneuverable themselves. Because the prey animals are usually smaller than the predators, the prey tend to have an edge and so pay a smaller evolutionary cost to become more maneuverable. All else being equal, a heavily targeted prey species should stay about even or slightly ahead of its most common predator in maneuverability. Fortunately for the predator, all else is rarely equal, and larger predators still have a substantial speed advantage.

Physics prevents large bombers from ever being as agile as contemporary fighters, although fly-by-wire controls and advanced structures give frontline bombers surprising maneuverability for their size. The U.S. Air Force's B-1B Lancer, about the size of a medium-large airliner, is reputedly as agile as much smaller airplanes, although when pilots say the B-1B "handles like a fighter," that is surely an exaggeration.[30] As good as they are, heavy bombers will never be able to use agility to avoid fighter attacks.

Fighters that attack other fighters are in an arms race comparable to that of aerial predators and their prey. Maneuverability is always at a premium, and one air force's improvements in fighter design tend to be matched or exceeded by opposing air forces. The continuing quest for ever greater maneuverability has led to fly-by-wire flight controls used in fighters so unstable that they are unflyable without computer assistance. Modern frontline fighters are all extremely fast and maneuverable, but that does not mean they are equal. Differences in acceleration and rate of climb, as well as maneuverability, can affect the outcome of a dogfight (a maneuvering air combat engagement). Moreover, which of two fighters is superior may depend on conditions. For example, early in World War II, the Japanese Zero was generally seen as superior to Allied fighters: at high altitudes, the Zero could outclimb and outturn them and was as fast or faster. Surprisingly, however, the U.S. P-40 Warhawk, usually described as obsolescent when the war started, was actually more maneuverable than the Zero at lower altitudes, particularly at high speeds.[31] Unfortunately for P-40 pilots, high-flying Japanese bombers meant that P-40 pilots usually fought under conditions that favored the Zeros.

Early Warning

For a flying animal in the open air, the best defense against aerial predators is to see the attacker early enough to take evasive action. The sky is a hard place to hide, so an alert animal should, in principle, be able to see an attacker from reasonably far away. If the potential victim sees the attacker in time, it may be able to dive for cover. For instance, if a duck sees a falcon stooping on another duck in the flock, the first duck can dive for the ground and hope to find shelter before the falcon turns its attention to other victims. If the falcon is too close to dive away from and if it is flying much faster than its intended prey, then the duck's best strategy is to watch and wait until the falcon is very close and then turn sharply (like a fighter evading a missile). If the duck gets the timing right, the falcon will overshoot, and the attacker's speed will prevent it from turning fast enough to intercept the duck.

Nocturnal insects have a different problem. At night, they cannot see well enough to visually evade bats (most can barely see that well in daylight), whereas the bat can easily "see" the insect using echolocation. Most insects, oddly enough, are deaf: they have no ears, no way to detect airborne sounds. But several different insects that fly at night, including praying mantises and a variety of moth species, have evolved ears that can hear bat sonar.[32] Most such insects have a sophisticated suite of evasive tactics that change as the bat gets closer. (Bats produce their echolocation cries faster and faster as they detect and then close in on an insect.) A moth might turn and fly at right angles to the bat's path when it first hears the bat, go into a diving turn as the bat gets closer, zig-zag irregularly if the bat gets closer still, and fold its wings and fall just before the bat intercepts it.[33] Some bats' sonar can actually detect insect wing beats; so if the insect stops flapping, the bat may mistake the now-falling insect for something inedible, such as a bit of leaf. Needless to say, bats have a much harder time catching insects that can hear their sonar than catching deaf insects.

Modern combat aircraft have an analogous capability provided by radar warning receivers, which "listen" for signals from other radars, whether on other airplanes or on the ground.[34] Such receivers typically show the type and distance to the other radars. Radar warning receivers can also inform the pilot, with increasingly strident alarms, that an enemy fighter has switched its radar from search to attack mode or has launched a radar-guided missile. A curious feature of radar warning receivers is that they can detect a radar signal much farther away than the radar itself can detect other airplanes. This is simple physics: for a radar set to detect another airplane, its signal must travel to the plane, be reflected, and travel back to the radar. The radar signal must travel twice as far to be detected by the radar set that produced it as it does to reach a radar warning receiver in the target airplane. So radar warning receivers will always "hear" radar well before the radar set actually "sees" the airplane carrying the radar warning receiver.

Why not just shut off your own radar and listen for the other guy's? In some situations that works; but as soon as the enemy figures out what you are doing, that pilot can do the same thing. Then both sides are back to depending on vision, not a favorable situation for jet fighters. Using a radar warning receiver alone (without radar) works best at high altitudes, when you have a chance of seeing someone sneaking up on you with their radar turned off and particularly if you can use ground-based radar or radar in a different airplane to warn you of the approach of an enemy with its radar turned off. Normally, radar warning receivers are used in conjunction with active radar, particularly to warn of rear attacks or ground-based attacks.

Active Countermeasures

Almost as soon as surface-based radar became common in World War II, researchers began to look for ways to interfere with it. One of the simplest is to "spoof" it. British researchers discovered that if airplanes drop strips of metal foil of the proper length, the foil acts as a strong radar reflector. If a plane drops lots of foil strips, the radar display becomes so cluttered that the airplane itself becomes difficult to pick out. British bombers soon began dropping great quantities of foil strips, code-named "window," to prevent German radar operators from getting a good radar image of the bombers.[35] Nowadays these strips are called chaff, and most combat aircraft carry a supply, some of which a pilot or a crewmember can eject when the radar warning receiver says that a targeting radar or radar-guided missile has locked onto their craft. Some airplanes carry a set of more sophisticated radar decoys that they can drop or tow. These decoys look like another plane on radar but take up more room than chaff bundles so fewer can be carried.[36]

Another technique, jamming, is simple in principle but difficult to apply effectively. When jamming, an operator broadcasts noisy radio signals on the same frequency as the radar signals to swamp returning radar echoes.[37] The basic problem with jamming is that the jammer is always faced with a choice: do I put all of my available power into a single frequency, or do I spread it out over a broad range of frequencies so my jamming is weaker on any one frequency?[38] The more power the jammer puts onto any one frequency, the more likely it is that the signal will swamp any radars using that frequency. Two or three decades ago, putting a lot of power into one frequency was effective against any radar set using that frequency. Nowadays, front-line military radar sets are frequency-agile, randomly switching from one frequency to another several times per second. The only way to jam such a radar set is to put a lot of jamming power into many different frequencies simultaneously. In practice, this means that jamming transmitters require so much power that they cannot be carried by fighter airplanes. Some bombers have limited radar jamming capabilities, but the U.S. military tends to combine radar jamming with other electronic countermeasures and place them on highly specialized, unarmed planes such as the Navy's EA-6B Prowler or the Air Force's EF-111 Raven. These planes travel with bombers as a sort of electronic escort, along with fighters that provide physical defense.[39]

One group of moths, the arctiid, or tiger, moths, do something remarkably similar to jamming. Not only can they hear bat echolocation calls, but they also have a structure near the base of each wing that produces ultrasonic clicks in approximately the same frequency range as bat sonar. When an echolocating bat closes in on a tiger moth, the moth begins emitting its own ultrasonic sounds, and in most cases the bat breaks off the attack. When biologists first discovered this ability, they assumed that the tiger moths were jamming the bats' sonar.[40] But some biologists soon began to question this idea. In order for the tiger moths to interfere with bat echolocation, the calls would have to be extremely loud (to swamp the real echoes) or very precisely timed and frequency-matched (to produce false echoes). An animal's sound power output is closely tied to body size, and a moth is probably too small to produce a loud enough sound to swamp a bat's sonar. So are tiger moth sounds emitted precisely enough to produce phantom echoes and confuse an attacking bat? Biologists have been arguing about this for more than a decade, but the consensus seems to be that moth sounds are not that precise.

If the moths are not jamming the bats' sonar, then what are their sounds for and why do they deter bat attacks? There are two likely possibilities: to produce a startle response or to act as a warning. Many insects, especially other butterflies and moths, depend on a startle effect for defense. Quite a few moths have large, bright, iridescent eyespots on their hindwings that they normally keep covered with drab, camouflaged forewings. If a foraging bird gets too close, the moth briefly uncovers the eyespots. Whether it is the sudden, unexpected appearance of bright objects or the eyespots' resemblance to the eyes of a predator such as an owl or a cat, flashing eyespots often appear to startle the bird and drive it off (see Chapter 2). Some biologists argue that tiger moth sounds affect bats in the same way that eyespots affect birds. They suggest that sudden sonar-like noises startle the bat and cause it to veer off to avoid them.[41]

Most biologists now favor the idea that the tiger moth sounds act as a warning. Many insects employ aposomatic, or conspicuous warning, colors: yellow and black stripes warn of a hornet's sting; orange and black wings warn potential predators that monarch butterflies are toxic. Like monarchs, tiger moth caterpillars accumulate toxins from the plants they eat; and not only are the adult moths extremely foul-tasting, but many also display aposomatic colors—orange and black or hot-pink and black—on their wings. The moths' clicks may do for bats at night what their bright colors do for birds during the day. According to biologist William Connor and his students at Wake Forest University, tiger moth sounds are better timed as warnings than as jamming.[42]

Maneuverability and Speed

Whether I am a fighter pilot or a falcon, if my task is to intercept another flying animal or an airplane on the wing, then I need some combination of high speed and superior maneuverability. Speed allows me to get close to another flyer, and maneuverability allows me to take quick action if the other flyer works to evade me. Yet at various times in the history of fighter development, designers and pilots have lost sight of the simultaneous importance of speed and maneuverability. Looking back at the success of highly maneuverable fighters such as the Fokker triplane of World War I, peacetime designers sometimes focused much more on maneuverability than speed, losing sight of the fact that the fastest fighters in the conflict were not all that much faster than the slowest fighters. Pilots encouraged this bias, probably because maneuverable fighters were more exciting to fly than faster, less maneuverable ones. In fact, Japan's famous Mitsubishi Zero, probably the best fighter in the Pacific at the outbreak of World War II, was initially rejected by active-duty pilots.[43] They were used to flying lighter, slower, but more nimble fighters and did not think the Zero was agile enough. Fortunately for them, their superiors overruled their objections and put the Zero into production.

Something similar happened in the United States between the world wars. Early Army and Navy fighters were state of the art and even competed successfully in air races in the 1920s. For example, Jimmy Doolittle first raced military planes as an Army pilot before leaving the service and becoming a successful civilian racer and test pilot.[44] Due at least partly to limited military budgets, U.S. fighter design stagnated in the late 1920s and early 1930s, and the few fighters that the services purchased emphasized maneuverability over speed (in spite of the "pursuit" designation of Army fighters). This stagnation led to a strange situation in the mid-1930s: civilian race planes and even some of the newest bombers became much faster than U.S. frontline fighters. When the prototype of the famous B-17 Flying Fortress bomber flew in 1935, it was more than sixty miles per hour faster than the U.S. Army's frontline fighter of the time, the P-26 Peashooter.[45] This may have been the factor that forced airplane manufacturers and the services to change their thinking and start designing high-speed, less nimble fighters. The large, fast, twin-engine Lockheed P-38 Lightning, which sacrificed some maneuverability for all-out speed, embodied this change in philosophy—and it was the first Allied fighter that was effective against the Zero.[46]

The P-38 was so sleek and fast that it was the first operational fighter to run into compressibility problems. *Compressibility* is an engineering term for changes in airflow patterns due to approaching or exceeding the speed of sound. Although the P-38 could not fly at the speed of sound, when its speed built up in a steep dive (during which the airplane as a whole might reach nearly two-thirds the speed of sound), the air flowing over certain parts of the airframe became supersonic, leading to control problems and structural failures. Because the whole airplane did not reach supersonic speeds, engineers did not immediately understand the problem and so could not initially figure out design changes to alleviate it.[47]

During the cold war, fighter design again emphasized only half the equation, except in this case speed won out over maneuverability. Jet engines replaced piston engines and the sound barrier was broken, so speed became the overriding feature of fighter design in the 1950s and 1960s. Military planners assumed that bombers would soon be supersonic so fighters would need to be even faster. The United States briefly employed a supersonic bomber, the B-58 Hustler, in the early 1960s and was also testing the prototype for an even faster bomber, the B-70 Valkyrie. Although the B-58 was phased out and the B-70 never went into production (partly due to cost and complexity, partly due to Soviet countermeasures), U.S. fighter designers continued to overemphasize speed because the Soviet Union had some supersonic bombers as well as some extremely fast fighters: their MiG-25 Foxbat, at more than Mach 2.5, was probably the fastest operational fighter ever produced. (Mach 2.5 is two and a half times the speed of sound, or more than 1,700 miles per hour).[48]

The U.S. Air Force and Navy initially fought in Vietnam with fighters designed (and pilots trained) to emphasize all-out speed, with sadly unimpressive results. This combat experience, in contrast to great success in Korea, led a number of senior officers to push for more maneuverable fighters, even at the expense of top speed. Although some fighters on both sides could fly at Mach 2 (more than 1,400 miles per hour), successful air combat never involved speeds faster than slightly over Mach 1. In fact, higher speeds just caused airplanes to run out of fuel during combat and get shot down. So for fighter airplanes, high speed may in fact have a practical upper limit.[49]

Aerial predators face almost the same demands as fighter airplanes for both speed and maneuverability. Dragonflies and insect-hawking bats are both extremely maneuverable for their size while also being much faster than their prey. Neither a dragonfly nor a bat can actually follow every twist and turn of an evading prey insect, but both are sufficiently nimble to correct for all but last-instant turns. Bats have the added advantage of a large prey-capture envelope: if they miss with their jaws, they can still make a capture using a wing or the tail membrane.

Peregrine falcons are a special case. They start so far above their target (hundreds of feet) and approach so fast that they almost ambush their target. Falcons use an amazing trick to accelerate at the beginning of a stoop. They often roll inverted so that their wings produce lift aimed downward. Their wings' lift adds to the force of gravity, giving them a downward acceleration that is faster than they would get from freefall. (Fighter pilots often start a dive the same way.)[50] Once it is speeding downhill toward the target, the falcon rolls back upright but with wings mostly folded so they produce little lift or drag, just control movements. Although some references claim that peregrines can reach more than two hundred miles per hour in a dive, this is an estimate rather than a rigorous measurement. The fastest diving speed accurately measured in a careful study by Duke University biologist Vance Tucker was 129 miles per hour. Tucker calculated that the falcon could have reached almost 160 miles per hour and still have leveled out successfully if it had not decelerated to land on its trainer's arm. This speed still makes falcons the fastest-known member of the animal kingdom.[51]

Nonetheless, reports still surface occasionally about supersonic bot flies, based on wildly inaccurate estimates published in 1927 by Charles Townsend. After a few years, during which the claim became entrenched in reference literature, Nobel Prize winner Irving Langmuir thoroughly debunked it. He pointed out that even if a bot fly could somehow summon the massive power needed to fly so fast, air pressure would immediately squash it. As of yet, no one has carefully measured bot fly flight speeds, but they are fast among insects and probably fly between twenty and thirty miles per hour, perhaps a bit faster. Some sources give sixty or ninety miles per hour as the top speed for insects, but I am skeptical of any reported speeds of more than twenty miles per hour.

The need for great maneuverability among both animal flyers and airplanes has led to a striking case of convergent evolution. The search for improved maneuvering led to aerodynamically unstable fighter designs that required active stabilization by the computers of a fly-by-wire control system (see Chapter 4). And in animals, unstable body shapes evolved along with actively stabilizing reflexes built into the nervous system.[52] Among animals, the victims, not the predators, have been at the front of the maneuverability arms race because they are under the strongest selection pressure: an unsuccessful predator temporarily goes hungry, but an unsuccessful prey dies. In aviation, however, fighters have led the maneuverability race, largely because high maneuverability is not an option for large bombers and transports.

In nature, predators have responded by evolving high maneuverability so they can continue to successfully intercept prey. In aviation, large airplanes have adopted fly-by-wire control systems for increased safety, versatility, and weight savings, not primarily to improve maneuverability. Still, the arms-race selection pressures that produced these active control systems in both animals and planes have been, if not identical, then surprisingly similar, even though changes in aviation technology are rapid and consciously goal-driven, while animal evolution is glacially slow, haphazard, and constrained by ancestry.

~ 10 ~

Biology Meets Technology Head On
Ornithopters and Human-Powered Flight

Machines that fly by either flapping their wings or using only muscle power make up a tiny, obscure corner of aviation technology. This niche may be outside the mainstream of aviation, but it has close ties to flying animals, especially birds. Ornithopters— machines that fly with flapping wings—are clearly inspired by animals, and muscle-powered airplanes use the same power source as animals do.

Why Link Ornithopters with Human-Powered Airplanes?
Ornithopters and human-powered airplanes have a deep historical association. Before the Wright brothers' flight (and, for a time, even after they flew), most people who were attempting powered flight built ornithopters. And because they lacked lightweight, powerful engines, the vast majority of those machines relied on the pilot to supply motive power. So ornithopters and human-powered aircraft are closely entwined in the early history of human flight. In fact, the two technologies did not diverge significantly until after World War I, long after the Wrights flew; and even then, ornithopters reappeared sporadically among human-powered airplanes for more than a decade.

Beginning with the earliest human attempts to fly and extending into modern scientific times, people have tried to emulate birds.

The earliest of these experimenters, whom aviation historians call "tower jumpers," generally followed the method of Daedalus and Icarus. The would-be flyer built a set of birdlike wings, often using actual bird feathers. He then climbed to the top of a tall building, strapped the wings onto his arms, and leaped. In spite of vigorous efforts to flap his wings, the tower jumper's inevitable fate was serious injury or death. Although tales of such attempts go back to the Roman Empire, reliable detailed descriptions are more recent. Recorded examples of such misguided dreamers include a Benedictine monk named Eilmer, who died trying to fly from a tower at Malmesbury in England in 1060, and John Damian, who attempted to fly from a tower near Edinburgh in about 1501 and escaped with only a broken leg.[1] Showing how little the state of the art had advanced, the Marquis de Bacqueville attempted to fly across the river Seine with wings strapped to his arms in 1742 and was seriously injured when he fell into a boat.[2]

As far as we know, Leonardo da Vinci was the first person to approach the problem of flight in a logical, scientific way. He designed machines to carry a person rather than just wings to strap onto a person's arms. Between 1480 and 1500, he drew designs for several human-powered ornithopters as well as helicopters.[3] In one example, he drew a clever transmission system to convert a rowing motion into vertical wing motion.[4] Nevertheless, his drawings were apparently just paper designs. Although some historians think he might have built one of these flying machines, no evidence that he did so has ever emerged, which is probably just as well. Given his available materials, none of his prototypes would have been light enough to actually fly under power.[5] Modern builders have built models and full-sized craft based on some of his designs; and though some have glided, none has achieved powered flight.

Aside from some tower jumpers and a few primitive attempts to build ornithopters, aeronautical progress after Leonardo stagnated until the beginning of the nineteenth century. In Vienna, clockmaker Jacob Degan built an ornithopter with strange, umbrella-like wings and a small hydrogen balloon to help support its weight and then managed to get his contraption a few feet off the ground in 1810. When he took it to Paris for a demonstration, the crowd was apparently so upset to discover that he had used a balloon that they severely damaged the craft. Degan's ornithopter was the earliest of many attempts in the first half of the nineteenth century to fly human-powered ornithopters, an interest probably spurred by the scientific and industrial revolutions then gathering steam in England and western Europe.[6]

By the mid-1800s, ornithopters had become more sophisticated and complex but no more successful. Between 1855 and 1868, French sea captain Jean-Marie le Bris built two ornithopters with boatlike bodies. They had large, flannel-covered wings that the pilot operated by pulling a set of cords. The cords operated levers through a pulley system to flap the wings. These craft did not fly under human power, and both were damaged in crashes when being towed by horses.[7]

Belgian Vincent de Groof built an ornithopter reminiscent of some of Leonardo's designs, with wings mounted on top of a telephone-booth-sized framework. The pilot stood in the framework and pulled on levers to flap the wings. De Groof could not convince authorities on the continent to let him try to fly there, so he took his machine to England. There, in 1874, he used a balloon to carry his machine up to a substantial height. When he cut the tether to the balloon, his flapping machine broke, and he fell to his death.[8]

English inventor Charles Spenser had a bit more success with his biplane-like ornithopter. The upper kitelike wing was fixed, and below it a pair of mobile wings were flapped by the pilot to provide thrust. Spenser apparently made some unpowered glides of distances up to 160 feet, but we have no evidence that he made any powered flights.[9]

Interestingly, Otto Lilienthal—the man who went on to develop and fly a successful series of gliders, to pioneer aeronautical research, and to inspire the Wright brothers—built an ornithopter long before he built any gliders. In 1867, he and his brother designed and built a six-winged ornithopter while they were still in school. They suspended their craft with a system of counterweights to test its drive system, but they never attempted free flights.[10] Many years later,

Lilienthal became a leading authority on flight and a highly success-ful glider designer. Like many others of his time, however, he contin-ued to believe that flapping propulsion was the road to aeronautical success. He was making plans to build an engine-powered ornithopter when he died in a glider accident.[11]

The success of the Wright brothers a few years after Lilienthal's death did not dampen the enthusiasm of ornithopter inventors. In 1908, both a Russian and a Briton tried to fly muscle-powered ornithopters without success. In 1920, Alexander Keith and José Weiss built a sophisticated machine with foot pedals for flapping the wings, which flew as a glider but did not fly successfully under power.[12]

Despite tremendous advances in fixed-wing, powered airplanes during World War I, some inventors persisted with ornithopters. In 1928, George White built a human-powered ornithopter in Florida that he launched from a speeding car. He claimed to have flown under muscle power, but he was never able to demonstrate muscle-powered flight in public demonstrations.[13] In Germany, a year later, Alexander Lippisch (who later became a famous aeronautical researcher) built a sailplane-like, pedal-powered ornithopter. Like so many of these machines, it made towed flights but never flew solely on muscle power.[14]

By the time White and Lippisch were building their machines, they had become lonely holdouts in a diverging trend between ornithopters and human-powered aircraft. The few people still experimenting with ornithopters had largely switched to engines for more power, recogniz-ing that human muscle alone was not powerful enough. Conversely, an equally small set of people working on human-powered aircraft had mostly adopted fixed-wing, propeller-driven designs because they real-ized that they needed all the efficiency they could get if they were to succeed with such limited power.

Humans As Power Plant

In spite of great advances in engine-powered airplanes throughout the twentieth century, a few visionaries kept plugging away at flight powered solely by human muscles. What seems blindingly obvious now, but what so many early pioneers overlooked or misunderstood, is that the mechanical performance of the human "engine" is the key factor that constrains all other design requirements. Before the 1930s, scientists had done few, if any, direct measurements of human power-production abilities. So designers of human-powered aircraft were stumbling in the dark when it came to the most fundamental design constraint on their machines.

In 1935, Oskar Ursinus established the Muskelflug Institut (Institute of Muscle-Powered Flight) in Frankfurt, Germany, with the goal of providing help and background information to human-powered aircraft designers. In addition to aerodynamic research, the Muskelflug Institut pioneered studies of human physical performance. Ursinus measured power outputs for various types of exercises—rowing, cranking, pedaling—by a variety of athletes and also measured how long they could sustain various levels of effort. He showed that his subjects could only produce their maximum power for about one minute, after which their power dropped by about 50 percent. He also found that, for short bursts, his subjects produced the most power if they could use both their arms and legs—for instance, by rowing or by simultaneously pedaling and cranking. But after two or three minutes, a subject's maximum effort using legs alone produced as much power as using both arms and legs.[15] World War II cut short the Muskelflug Institut's research, however, and the study of human muscle performance languished for the next couple of decades.

In the late 1950s, some members of the Royal Aeronautical Society became interested in human-powered flight. They quickly realized that they needed data on the "engine" as a basic ingredient in the design of an airplane. They approached Douglas Wilkie, a well-known muscle physiologist, for information. When he agreed to help them, as he wrote in an essay some twenty-five years later, he expected that he would only need "a few afternoons in the library" to satisfy the request.[16] Instead, he ended up performing a detailed study of exercise duration and power output, examining both elite athletes and average but highly trained undergraduates. His research has become the benchmark for studies of human performance and the starting point for the design of all modern human-powered aircraft.

Wilkie found that Olympic athletes produced significantly more power than non-athletes, both in maximum performance bursts and over longer durations. What surprised him was that this difference persisted even after the non-athletes had undergone months of intensive exercise training. (In a bit of scientific understatement, Wilkie called his highly trained non-athletes "average fit men," which, when used out of context, has occasionally misled people into thinking his data apply to *untrained* individuals.) Wilkie found that the strongest humans can produce about 1.3 horsepower (approximately 1,000 watts) in brief bursts of a few seconds; but as exercise periods last longer and longer, both maximum and sustainable power drop off dramatically. For exercise lasting longer than three minutes, elite athletes can sustain the production of slightly more than 0.5 horsepower (approximately 400 watts). Wilkie's data for the maximum short-term power output for elite athletes are similar to that of Ursinus, but his measurements on non-athletes were the first of their kind.[17]

The difference between power output during brief bursts and more sustained exercise comes about because of the different ways in which muscle can burn fuel as well as different fuel types. Muscles maintain a store of carbohydrates (derived from sugar), which they can use to power muscle contraction anaerobically—that is, without oxygen. Using carbohydrates anaerobically is inefficient and produces lactic acid as a byproduct. Anaerobic carbohydrate consumption is fast, however, and produces a quick burst of high power; but once the store of carbohydrates is consumed, that source of power is exhausted. Muscles can also use fuel carried in the blood, primarily fats, which muscles consume aerobically, using oxygen. This aerobic process is more efficient—transferring more energy to the muscles for each ounce of fuel—and produces carbon dioxide rather than lactic acid as a byproduct. The power output of muscles working aerobically is limited by how fast the body can supply the muscle with oxygen, and the maximum amount of aerobic power is typically half or less than the maximum anaerobic power. The aerobic power level, however, can be sustained for many minutes, even hours, depending on the body's fuel reserves. We are intuitively familiar with these muscle properties: we expect a sprinter's speed over one hundred yards to be much faster than a marathon runner's speed over twenty-six miles. The sprinter's muscles work anaerobically on stored carbohydrates, whereas the marathoner's muscles work aerobically and need the blood to provide a continuous oxygen supply.[18]

The early human-powered flight pioneers were undoubtedly aware that people can produce more power for short bursts than for long periods, but they tended to overestimate the maximum power a pilot could produce, even for short bursts. The failure of human-powered flight experimenters between the early flights of the Wright brothers and World War II was due as much to incorrect assumptions about human power output as to aerodynamic problems.

Modern Human-Powered Flying Machines

From 1912 though the 1920s, the French passion for all things aeronautical included human-powered aircraft. French enthusiasts developed a type of human-powered airplane in this period called the *aviette*. Aviettes were basically winged bicycles. Most were powered only by their bicycle wheels on the ground; once the drive wheel left the surface, they became gliders. Few had any system of control, which was fine because not many actually left the ground and the ones that did were airborne for such a short time that no maneuvering was necessary. In 1912, M. Didier launched his aviette off a short ramp and was airborne for about sixteen feet, a distance that may not have even required wings. By 1921, Gabriel Poulain was making glides of forty feet with his aviette. Although some aviettes had propellers (or even flapped their wings) to assist the drive wheels, most, including all those that actually flew, were strictly wheel-driven. These machines were really muscle-launched gliders rather than true human-powered aircraft.[19]

The restrictions on powered aviation that stimulated soaring and gliding in Germany after World War I also prompted interest in human-powered aircraft. We have already encountered Lippisch's

1929 muscle-powered ornithopter, and inventors in Germany designed several other wing-flapping and propeller-driven human-powered aircraft in the 1930s. Of the few that were actually built, only one achieved successful human-powered flight. This was the Haessler-Villinger craft, the first really successful, modern human-powered aircraft (HPA).

Helmut Haessler and Franz Villinger built their machine in Germany in the 1930s. It looked like a typical sailplane of the period except that it had a propeller on a tall pylon in front of the wing and was much lighter than a conventional sailplane (fig. 10.1). They called their craft Mufli, a name that looks suspiciously like a contraction of the German words for muscle (*muskel*) and flyer (*flieger*). Mufli had a wingspan of more than forty-four feet and weighed only seventy-five

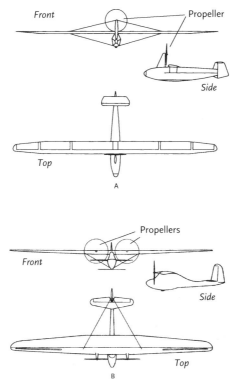

Figure 10.1   Three-view drawings of (A) Haessler and Villinger's Mufli (top) and (B) Bossi and Bonomi's Pedaliante.

pounds without its pilot. In late August 1935, Mufli's first series of flights reached a maximum distance of 720 feet with a glider pilot at the controls. When a professional bicycle racer replaced the glider pilot, Mufli flew more than 1,300 feet on several occasions. In 1937, the same pilot flew Mufli a distance of 2,336 feet, a record that stood for more than two decades.[20]

In Italy, at about the same time, Enea Bossi and Vittorio Bonomi developed a craft they called Pedaliante. It also looked rather like a contemporary sailplane except that it had two propellers on the front of its wing (fig. 10.1). Pedaliante had a wingspan of fifty-eight feet and weighed a hefty 220 pounds empty. (Bossi claimed that at least thirty kilograms, or sixty-six pounds, could be chalked up to the Italian government's airworthiness requirements for standard airplane structures.) The pilot's pedals drove a large bicycle wheel as well as the two propellers. In 1936, Pedaliante flew three hundred feet on its first entirely muscle-powered flight. It eventually flew a full kilometer (more than 3,000 feet) after a catapult launch, but Bossi's pilots had trouble even repeating the three-hundred-foot flight without a catapult launch.[21]

World War II put an end to experiments in human-powered flight, and progress did not really pick up again until the 1960s. Sadly, both Mufli and Pedaliante were destroyed during the war.

Postwar Progress

In the years immediately following World War II, most people in Great Britain and continental Europe were too busy recovering from the war to put effort into something as frivolous as human-powered flight. In 1957, however, a group of enthusiasts formed a committee to promote human-powered flight efforts at Cranfield College of Aeronautics, an elite graduate engineering school established by the Royal Air Force. Committee members initiated limited studies on human-powered aircraft designs; but as committee member Beverly Shenstone (chief engineer of British European Airways) noted, research could not progress without a machine to test the theories. The next year, this committee was folded into the Man Powered Aircraft Group of the Royal Aeronautical Society (RAeS).[22]

Committee members were probably not expecting practical applications for human-powered flight. I suspect that they were motivated by the technological challenge of the endeavor. From comments made by RAeS group members in connection with some of the later prize competitions, at least some group members hoped that human-powered aviation would take off as a sport, much like soaring.

The Man Powered Aircraft Group of the RAeS began raising funds to support research into HPAs and to support building of human-powered airplanes. The RAeS group did three important things to stimulate interest and progress in human-powered aircraft. First, they were the ones who approached Douglas Wilkie for information on human muscle performance. Wilkie's information gave human-powered aircraft designers more detailed and accurate information on their engines than they had previously had.[23] Second, they raised money and provided small grants to people interested in building human-powered airplanes, which funded research, design, and construction. Third, and probably most important, they worked with Henry Kremer to set up the Kremer Prize.[24]

Henry Kremer was a wealthy engineer and businessman who directed several industrial companies. He had developed a number of industrial processes, including one for making the plywood used to build the World War II de Havilland Mosquito fighter-bomber. He was also an aviation buff, a physical-fitness enthusiast, and a close friend of Robert Graham, first chairman of the RAeS Man Powered Aircraft Group. In 1959, responding to a suggestion from Graham, Kremer agreed to donate 5,000 pounds sterling as the prize in a human-powered aircraft competition to be administered by the RAeS.[25] (Although 5,000 pounds, worth about 14,000 dollars at the time, may not sound like an enormous sum today, it was then equivalent to the full purchase price of a medium-sized house.) The Man Powered Aircraft Group used Wilkie's data on power and duration to devise rules for a contest that they judged to be just barely feasible. Announced in 1960, the rules required the winner to take off and fly solely under muscle power, fly a figure-eight course around two markers one-half mile apart, and cross the start and finish line at an altitude of at least ten feet. The initial competition was open only to citizens of British Commonwealth countries.[26]

The RAeS committee was remarkably effective in jump-starting developments in human-powered flight in England. Not only did the Kremer Prize stimulate interest in human-powered airplanes, but it inspired others to contribute money to the RAeS to fund their HPA grant program. The first two HPAs to fly in Great Britain flew within a week of each other in 1961. The first was SUMPAC (an acronym for "Southampton University Man Powered Aircraft"), followed a week later by Puffin. On November 9, SUMPAC made a flight of 210 feet. The craft was designed and built by a team of college students led by Alan Lassiere, Anne Marsden, and David Williams (fig. 10.2). The team leaders began the project while still undergraduates in aeronautical engineering at Southampton University, and they completed the craft with the assistance of the university's Department of Aeronautics and a 1,500-pound grant from the RAeS. The machine

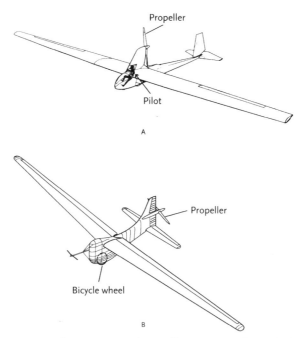

Figure 10.2 Drawings of (A) SUMPAC and (B) Puffin I. SUMPAC shows pilot location and a propeller mounted on pylon. Note Puffin's tail-mounted propeller. Both have a bicycle wheel as the main landing gear (here shown only for Puffin).

was enormous for its weight: with a wingspan of eighty feet, it weighed 128 pounds without the pilot. Although it weighed a bit more than Mufli's seventy-five pounds, it also had more than three times Mufli's wing area.[27]

A week after SUMPAC's first flight, Puffin flew successfully. This craft was built by a group of enthusiasts, the Hatfield Man Powered Aircraft Club, most of whom worked at the de Havilland Aircraft Company factory in Hatfield. Puffin was similar in size to SUMPAC, with a wingspan of eighty feet and an empty weight of 118 pounds. Both SUMPAC and Puffin had much longer wings than either Mufli or Pedaliante, and they were little more than half the weight of the latter.[28]

The structure of SUMPAC and Puffin had some similarities and some significant differences. They both had a large bicycle wheel as well as the propeller connected to the pilot's pedals to give added acceleration while on the ground. In SUMPAC, the pilot leaned back in a semi-reclined seat behind the wheel, but in Puffin, the pilot leaned forward above and in front of the drive wheel. Both had wings built out of balsa and spruce formed into complex arrangements of trusslike ribs and spars, with the front of the wing covered by thin balsa sheeting and the rear covered by fabric or plastic film. SUMPAC started out mostly covered by nylon parachute cloth, but the cloth tended to sag when the weather changed, so the builders eventually replaced it with a covering of plastic film like Puffin had (Melinex, similar to Mylar). SUMPAC looked somewhat more conventional, rather like a sailplane with a pusher propeller on a pylon over the wing, a similar layout to Mufli's. Puffin, designed by professional engineers and largely built by apprentices at de Havilland's technical school, looked smoother and more refined, although it had a rather blunt, bulbous nose (fig. 10.2). Puffin's propeller was mounted on its tail and was driven by a long shaft linked to a sophisticated gearbox connected to the pilot's pedals.[29]

Although SUMPAC flew first, its builders had limited resources (they had to store the craft in a damp, leaky Quonset hut at an airport twenty-five miles from their school), and group members tended to disappear as they graduated and left for jobs. They recruited gliding instructor Derek Piggot as pilot, and he trained intensively to come up to the standard of Wilkie's "average fit man." The Hatfield team, in contrast, had the backing of the de Havilland Aircraft Company, including the use of a hangar and a runway, plus the services of de Havilland test pilots. Despite a series of incremental improvements to SUMPAC, the Southampton students did not have the time or the resources to outperform the Hatfield group. By late 1962, both crafts had made long flights. SUMPAC achieved flights of nearly 2,000 feet, and Puffin set a straight-line distance record of 2,980 feet that stood for ten years. Turns, however, proved elusive for both, mainly because flight durations were so limited and exhausting that the pilots did not have time to do any maneuvering. Puffin eventually made unassisted turns of up to eighty degrees, but SUMPAC was difficult to control and barely managed to fly straight. Both teams turned to model airplane engines to give their pilots a break from pedaling and let them concentrate on turns. With a model airplane engine for power, SUMPAC finally made an 80-degree turn, and Puffin actually made an engine-assisted flight with one 270-degree turn followed by a 90-degree turn the other way. But neither aircraft had anywhere near the endurance or the turning ability to fly the Kremer course on muscle power.[30]

Alan Lassiere took SUMPAC with him when he moved to London for a new job. He intended to modify and improve it; but unfortunately, after two years of rebuilding, it crashed in a gust of wind before it could be fully tested and never flew again.[31] Puffin met a similar fate in 1963: a gust blew it off the runway into deep mud, causing a sudden stop that destroyed its wings. The craft was insured for 4,500 pounds, however, so the team used the insurance proceeds and the old fuselage and tail of Puffin to build Puffin II. Puffin II had a longer wing and, for reasons never clearly explained, was more than twenty pounds heavier than Puffin. But the bigger wing and heavier weight more or less canceled each other, so Puffin II required essentially the same amount of power to fly as the original craft had.[32]

News of the Kremer Prize and the successful human-powered flights in England stimulated Professor Hidemasa Kimura to

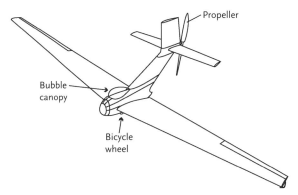

Figure 10.3 Nihon University's Linnet II. Its main difference from Linnet I was its upright pilot seating, which required a bubble canopy. Linnet I's pilot reclined and was entirely contained in the upswept fuselage.

challenge his aeronautical engineering students to build a human-powered airplane, even though they were not eligible for the original Kremer Prize. They initiated a long-term, multiyear project at Nihon University in Japan, which produced a series of sleek, lightweight, and increasingly capable aircraft. The first of these, Linnet I, flew a modest fifty feet in 1966, but Linnet II flew three hundred feet in 1967 (fig. 10.3). These machines led to the Egret and record-setting Stork series aircraft of the 1970s.[33]

Back in the United Kingdom, Puffin II proved harder to control than Puffin had been and never managed to fly as far. Another craft, Daniel Perkins's ingenious, inflatable Reluctant Phoenix, made several flights in an ex-dirigible hangar. With a longest flight of more than four hundred feet, Reluctant Phoenix was the third HPA to fly successfully in the United Kingdom, but Perkins abandoned the project when he realized that he had no chance of flying the Kremer course.[34]

Next Generation

In 1967, the RAeS and Henry Kremer decided to up the ante. The original Kremer Prize was about to expire; so Kremer increased the prize amount to 10,000 pounds, and the RAeS opened the competition to competitors from any country.[35] No doubt inspired by the

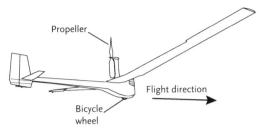

Figure 10.4 Jupiter in flight.

larger prize but also building on the limited success of earlier human-powered airplanes, a second generation of HPAs took flight in the United Kingdom during the 1970s.

In the mid-1960s, Chris Roper had started an HPA group that set about building a craft based on the best features of SUMPAC and Puffin. It had the same basic layout as SUMPAC, with the propeller on a pylon over the wing and a pedal-driven bicycle wheel for ground acceleration. The craft, called Jupiter, was badly damaged in a fire in 1969 before it was finished, at which point John Potter took over the project. Potter was a flight lieutenant in the Royal Air Force and head of the RAF Apprentice Training School. He and his students repaired and completed Jupiter in 1971 (fig. 10.4); and Potter, after intense conditioning, became its pilot. During the next four and a half months, he flew Jupiter sixty-four times, accumulating about an hour of total flight time, far more than any other HPA had achieved to that date. In 1972, Potter flew Jupiter more than 3,500 feet, finally breaking Puffin's 1962 record.[36]

A craft sponsored by the British Aircraft Corporation was also completed and flown in 1971. Called Dumbo, it used sophisticated construction techniques—chemically milled aluminum tubing with fiberglass joints—to produce a very large but light craft. Despite an enormous 120-foot wingspan (larger than a Boeing 737's), Dumbo weighed only 178 pounds empty. It had the usual powered bicycle wheel and a tail-mounted propeller like Puffin's. Its name arose from the fact that it was steered by differentially tilting the entire right and left wings, which reminded team members of the Disney cartoon

elephant who flew by flapping his outsized ears. Despite its sophistication, however, Dumbo had serious control problems, and its
longest flight was a disappointing 150 feet.[37]

A number of other human-powered airplanes made successful
flights in the early 1970s, including Toucan, a two-person craft, in
England ("toucan fly if one cannot") and Aviette in France. Named in
honor of the winged bicycles of previous decades, Aviette flew more
than 3,000 feet but shared the turning problems of earlier craft. None
had the duration or control needed to fly the Kremer course, which
was their primary goal.[38]

In 1973, probably wondering if his prize would ever be won, Henry
Kremer increased the prize amount to 50,000 pounds (slightly less
than the equivalent of 130,000 dollars at the time).[39] This increase
made it the largest aviation prize ever offered and seems to have finally
sparked some U.S. interest in human-powered flight. In 1976, Joseph
Zinno made the first successful human-powered flight in the United
States in his Olympian ZB-1. Operating basically as a one-man show,
Zinno, a retired Air Force officer, built his craft alone and flew it
himself. He made one flight of seventy-seven feet, apparently the only
time the Olympian flew under muscle power.[40]

Paul MacCready and the Gossamers

In 1976, Californian Paul MacCready—champion model airplane
builder, champion sailplane pilot, engineer, hang glider pilot, and
inventor of the MacCready speed ring (see Chapter 7)—was looking
for a change. He had started or purchased several companies, some of
which had been modestly successful, one of which had folded and left
him with a 90,000-dollar debt. He had just written an article on hang
glider performance and, on an extended cross-country vacation, had
become fascinated by soaring birds and how they compared with
hang gliders. At some point on that trip, he started thinking about
human-powered flight. He gradually came to the conclusion that if he
could scale up a hang glider without increasing its weight, he could
fly it on about 0.3 horsepower (220 watts). Three-tenths of a horsepower is a power output that a person can produce.

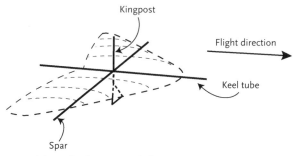

Figure 10.5 Hang glider with a kingpost-style framework (wing surface shown as dashed). Bracing wires run from both ends of each frame tube—kingpost, keel tube, and spar—to the tips of the other frame tubes (wires omitted here for clarity).

MacCready decided he wanted to take a shot at winning the Kremer Prize.[41] Yet he took a radically different approach from that of previous contenders. He started with two fundamental design features. First, he decided the craft needed an extremely long, lightweight wing. If he could keep the craft light enough, it would fly slowly and allow him to use his second basic feature: a wire-braced kingpost framework. Most hang gliders at that time used such a framework, a series of struts arranged like one of the pieces in the game of jacks (fig. 10.5). The kingpost itself is a vertical tube, which is connected to two horizontal tubes. The horizontal tube running right and left forms the main wing spar, and another horizontal tube running fore and aft is called the keel tube (or just the keel). By stringing bracing wires from the top and bottom of the kingpost to the ends of the horizontal tubes, he produced a very strong, lightweight structure.[42] Other HPA builders would probably have been appalled at these drag-producing bracing wires, but MacCready had a crucial insight: because the wires' drag is proportional to the square of the airspeed, the lower structural weight would more than make up for the added drag of the wires if he could fly slowly enough.[43]

MacCready soon settled on a radical layout to go with his radical structure: an enormous, ninety-five-foot wing, initially with a single surface—that is, ribs with covering only on top. The horizontal stabilizer was mounted on the *front* end of the keel tube, making it a canard ("tail first") design. The prototype had vertical stabilizers on its wingtips (later abandoned), the propeller behind the wing at the back end of the keel tube, and no fuselage as such. A seat, pedals, and controls for the pilot were mounted on the lower end of the kingpost, under the wing.

With his family and a few friends, MacCready built the prototype in a couple of weeks in a pavilion used to build floats for the Rose Bowl Parade in Pasadena, California. Accounts differ about whether it actually flew before being badly damaged by a gust of wind on a test glide in the Rose Bowl parking lot. MacCready then found hangar space at Mojave Airport, near Pasadena, and in the fall of 1976 set about building a team and testing and refining his craft. The team called the machine the Gossamer Condor.[44]

Early on, MacCready made several important decisions. First, construction would be "quick and dirty," with all structures easy to repair. (Eventually, the team adopted the slogan, "If it hasn't broken yet, it is too strong.") Second, he gave up on flying in ground effect (a reduction in induced drag when flying very close to the ground), figuring that the structural advantages of a wing mounted high on a kingpost were more important than working out a way to mount the wing down low. He also believed this arrangement might make turns easier to achieve.[45] Finally, he dispensed with the pedal-driven bicycle wheel. He decided that if the craft could fly and turn just with the propeller, then it certainly ought to be able to take off that way, particularly if he could keep down the weight and the speed. Instead of a bicycle wheel, the Gossamer Condor used a fore-and-aft pair of toy truck wheels for landing gear, purchased in bulk and treated as disposable.[46]

MacCready had originally planned to use composites for the framework but was advised by an expert to stick with aluminum tubing that had been chemically milled to decrease its wall thickness. A team member described the wing spar as equivalent to the diameter and strength of an aluminum beer can but ninety-five feet long and braced with steel wires.[47] The first version of the Gossamer Condor looked like little more than a huge wing, with another little wing hung out in front and a stripped-down bicycle frame hanging below. Even with its wingspan increased to ninety-six feet and innumerable repairs and refinements, the final version weighed only seventy pounds.[48]

A New Beginning

After a frustrating period at Mojave Airport, where inconsistent flight distances, poor control, and limited test opportunities due to high winds plagued the craft, MacCready decided the Gossamer Condor needed a major redesign and a new home. In February 1977, the team moved to Shafter Airport in the San Joachin Valley.[49] Team members built a redesigned, tapered, swept, double-surface wing (covered top and bottom) and a fully enclosed cockpit. The whole craft was covered in Mylar plastic film (fig. 10.6). On March 4, one of MacCready's sons

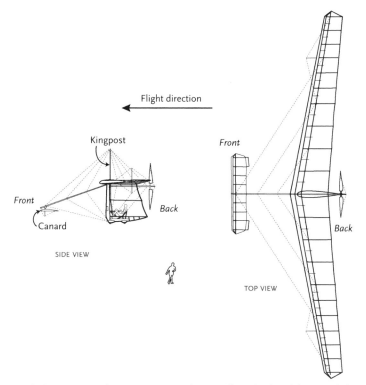

Figure 10.6 Gossamer Condor structure. Note the propeller in back and the canard elevator in front. Bracing wires are shown as dotted lines, with only a subset drawn here for clarity. The human figure is shown for scale.

test-flew the new version and found that it was much easier to fly but still difficult to turn. Even so, on March 6, bicycle racer Greg Miller flew it for more than five minutes, the longest flight duration of any HPA to that time. This flight did not set a distance record because the craft flew into a four-mile-per-hour headwind and was much slower than other HPAs. Even so, its distance over the ground was just under 2,000 feet, and its distance through the air was probably more than 3,000 feet. But team members still had not figured out how to safely and reliably turn the craft.[50]

The problem with turning any HPA is that these craft fly so slowly and their wings are so long that, in a turn, the airspeed of the wing toward the inside of the turn drops enough to lose lift. This loss of lift not only prevents the pilot from leveling back out but may cause an uncontrolled increase in bank angle and lead to a crash. MacCready's original control concept for the Gossamer Condor was to use the canard (front horizontal stabilizer) as an elevator, tilting it up and down to control pitch, and spoilers near the tip of each wing to control roll. But given the marginal engine power available, lift-killing and drag-producing spoilers seem to be an odd choice for controlling a human-powered airplane. His team quickly abandoned them and came up with an ingenious idea: why not bank the canard and let it tow the main wing around the turn? This worked better than the spoilers but was still not effective and consistent enough.[51]

Henry Jex, an aerodynamicist who was helping MacCready with control analysis, recommended wing warping—that is, twisting down the trailing edge of the outer part of one wing like an aileron, as the Wright brothers did on their original airplanes.[52] MacCready was initially skeptical; but when team members eventually tried it, they discovered something startling. Wing warping was very effective, but it worked opposite to how normal ailerons functioned. On a conventional airplane, deflecting the right aileron down increases the lift on the right wing and raises it, banking the plane to the left. When the trailing edge of the Gossamer Condor's huge wing twisted down, the wing produced a bit more lift, but its inertia was so great that it didn't bank to the left. Instead, the increased drag of the deflected trailing edge yawed the wing to the right, and the extra lift compensated for the decrease in speed of the inner wing and prevented overbanking. In essence, the Gossamer Condor turned using adverse yaw (see Chapter 4).[53]

Gossamer Condor pilots found that they could make fully controlled turns by reverse wing warping to get the craft to swing to one side and banking the canard to fine-tune the turn. On April 5, they made the first successful 180-degree turn using this system. By April 26, they had accumulated more than three hundred flights on various versions of the Gossamer Condor, more than all other contemporary Kremer contenders combined.[54] Working throughout the summer, flying at dawn when the winds were calmest, they continually refined the craft. In August, hang glider pilot and bicycle racer Bryan Allen made an unofficial test flight that followed the entire figure-eight Kremer course. Although downdrafts caused the nose wheel to touch the ground a couple of times, the team was confident that Allen would be able to fly the course for the prize. On August 23, he successfully flew the Kremer course for the prize, with his total flight lasting seven minutes and twenty-five seconds and covering approximately 1.35 miles. The time between crossing the start and finish lines was six minutes and twenty-three seconds for a distance of 1.15 miles at approximately eleven miles per hour.[55]

There are several explanations for why Paul MacCready's group succeeded so quickly when so many others had failed after prolonged work. First, a very large team worked on the craft, with several people working more than full time. Second, because of the craft's easy-to-repair design, what might have been a project-ending crash for another craft required only a couple of days of repair for the Gossamer Condor. Finally, the Shafter location had consistently good weather, so the team rapidly accumulated more flight experience in a few months than other teams had achieved over the course of years.

Channel Challenge

Two weeks after the Gossamer Condor's prize-winning flight, Henry Kremer announced a new prize of 100,000 pounds (equivalent to about 170,000 dollars at the time it was offered) for the first HPA to cross

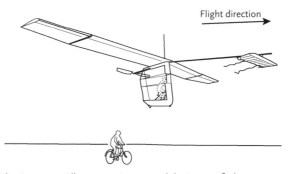

Figure 10.7 The Gossamer Albatross as it appeared during test flights.

the English Channel. MacCready and some of his team decided to try for that prize as well. They designed a sleeker, more high-tech craft they called the Gossamer Albatross. This machine used the same basic layout as the Condor but was more refined (fig. 10.7). Its frame was built of custom-made, carbon fiber–epoxy composite tubes and was rigged with nylon and Kevlar lines instead of steel wires. Aside from the pedals and sprockets of the drive system, almost no metal was used. Even tiny control-cable pulleys were made of special plastics.[56]

The Gossamer Albatross first flew on July 22, 1978, less than a year after the Gossamer Condor had won the first Kremer prize. Unlike the Gossamer Condor, the Gossamer Albatross flew reasonably well from the start. The pilots found that it had excessively sensitive controls, however, which caused one serious crash after just a couple of weeks of testing.

MacCready's biggest problem was sponsorship. The expenses for the cross-channel attempt were enormous, including insurance, transport of people and equipment, chartering of safety boats, and lodging for team members while waiting indefinitely for suitable weather. After multiple pleas, the manufacturer Du Pont agreed to sponsor the team, a step that was eminently appropriate. After all, Du Pont had made the Mylar covering, the Kevlar control lines, and most of the plastics used to make pulleys, fairleads, sprockets, and so on.[57] Du Pont balked, however, at covering the cost of transporting the Gossamer Albatross and two backup planes to England by air freight. But as a result of a fortuitous series of conversations with a Navy test pilot, MacCready was able to arrange to piggyback transport of his airplanes onto a training flight of an RAF cargo plane scheduled to leave from Nellis Air Force Base (near Las Vegas) at the end of April 1979.[58]

At the time of MacCready's deal with the RAF, the Gossamer Albatross had not made any flights much longer than the Gossamer Condor's. Just a week before the scheduled departure, the team installed a new, more efficient propeller. This improved the Albatross's performance enough for Bryan Allen to make a flight of more than sixty-nine minutes, by far the longest human-powered flight to that date. Soon after, the Gossamer Albatross was disassembled and crated for shipment.[59]

Once in England, the team had to overcome many organizational hurdles: coordinating with the British and French coast guards; finding hangar space, work space, lodging, boats, radios, and navigation equipment. They also faced two months of waiting for the weather to cooperate.[60] MacCready's funding was about to run out when a break in the weather gave them a chance to attempt the flight. On June 12, 1979, they assembled the Gossamer Albatross on the waterfront near Folkestone, and Bryan Allen took off. Fighting headwinds, unexpected turbulence, leg cramps, and dangerously inconsiderate reporters' boats, Allen pedaled the craft for two hours and forty-nine minutes to fly twenty-two miles from the British coast to a French beach. He almost gave up partway across the channel; but when he climbed up a few feet to hook onto a pole to get a tow from one of the chase boats, he found smoother air and easier pedaling. After two hours, the batteries in his altimeter (modified from the sonar range finder of a Polaroid Swinger camera, discussed in Chapter 6) and airspeed indicator died, and he ran out of drinking water. These were not trivial problems. As seaplane pilots know, judging height over water from an airplane is very difficult. Moreover, Allen perspires more than one quart of water per hour when he pedals at peak output, and his leg cramps were undoubtedly due to dehydration. Displaying unbelievable endurance, he was within a few seconds of his previously measured, absolute maximum physiological performance limit when he touched down on the coast of France.[61]

Other HPA Milestones

Even before the Gossamer Albatross won the Cross Channel Prize, Henry Kremer and the RAeS announced other prizes. One, the Kremer Figure-Eight Prize, promised to award 10,000 pounds to the second human-powered airplane to fly the standard Kremer figure-eight course, with the caveat that the craft be from a country other than that of the winner of the first Kremer Prize. And soon after the cross-channel flight, Kremer and the RAeS announced the Kremer Speed Prizes. The first of those prizes, with a purse of 20,000 pounds, would go to the first HPA to fly around a triangular, one-mile course in under three minutes. Succeeding 5,000-pound prizes would be awarded to craft that could beat previous records by at least 5 percent. Energy storage devices were allowed as long as they used energy produced by the pilot during a restricted period immediately before the flight.[62]

Probably the most successful of several craft built soon after the Gossamer Albatross were the Musculair I and II, designed and constructed by a small German team led by Gunther Rochelt and flown by Gunther's son, Holger Rochelt. In June 1984, Holger pedaled Musculair I around the Kremer figure-eight course, making it the first craft built by someone other than MacCready to do so. As we shall see, Musculair I also went on to win one of the Kremer Speed Prizes. And it also became the first passenger-carrying HPA in history when the Rochelts taped a makeshift seat onto the frame tube behind the pilot. Holger's young sister climbed onto the passenger seat, and Holger piloted her on a flight of just under 1,700 feet. After Muscular I was damaged in a transport accident, the Rochelts built a smoother, more refined version, Musculair II. This craft won the fifth and final Kremer Speed Prize, beating the previous time by nineteen seconds, a whopping 13 percent improvement.[63]

The Kremer speed competition involved two other veteran teams, the MacCready team and a group of students and former students at the Massachusetts Institute of Technology (MIT). The MIT team had built two HPAs before the speed competition: one successful, one not. For this competition, MacCready's team members built a slightly more conventional-looking craft than their previous designs had been, with a conventional aft tail on a long boom, which they dubbed the Bionic Bat. The MIT group produced a craft with a similar layout, which team members called Monarch.

Although the Bionic Bat flew the prize course in less than three minutes in September 1983, its claim was disallowed due to technical issues surrounding the team's electric energy storage system. The Monarch won the first Kremer Speed Prize in May 1984, and the Bionic Bat won the second prize two months later. Both craft used electric energy storage systems that the pilot was allowed to charge up by pedaling for ten minutes before the flight. In contrast, Musculair I, winner of the third speed prize, and Musculair II, winner of the fifth speed prize, did not use any type of energy storage, relying entirely on the pilot's muscles in flight. Even so, Musculair II convincingly smashed the speed records of all previous winners.[64]

After the RAeS closed the Kremer speed competition, HPA enthusiasts wondered what could be next for human-powered flight. Some members of the Monarch team proposed re-creating Daedalus's mythical escape from Crete by flying across the Aegean Sea from Crete to the nearest landfall. Their goal was the island of Santorini, more than seventy miles away.[65] Several team members formed the nucleus of the new group that would attempt this long overwater flight.

Led by John Langford, they decided to proceed very carefully and methodically rather than in the rush needed to compete for the Kremer prizes. They first built the Michelob Light Eagle (sponsored by the Anheuser-Busch Brewing Company) as a prototype to test their design. This craft flew at the NASA Dryden research facility near Edwards Air Force Base and set a new human-powered distance record of 36 miles.[66] Although the Eagle's record was more than double the previous HPA distance record, it was only half the distance between Crete and Santorini. Eventually, the team accumulated more than a half-million dollars in sponsorship funding for two definitive craft, dubbed Daedalus 87 and Daedalus 88. They also recruited champion cyclists and trained them to fly—first in sailplanes, then in Daedalus simulators—rather than trying to find pilots who also happened to be champion bicycle racers.[67]

The Michelob Light Eagle and the two Daedalus craft all had sleek, clean designs, resembling small futuristic airplanes rather than the unconventional, wire-braced, frail-looking Gossamer Albatross. In spite of their neat, high-tech appearance, however, the two Daedalus airplanes were every bit as fragile as the Gossamer craft: they had to be delicate in order to be light enough for their epic flight. One curious design feature was that, to reduce weight, they had no ailerons or wing-warping mechanism. Instead, they turned using the rudder, as if they were enormous radio-control models.[68]

Both Daedalus 87 and Daedalus 88 had flown by early 1988, and design improvements suggested by the former were built into the latter. In March, the Greek air force transported both the Daedalus machines as well as the Eagle to Crete, free of charge. The Greek force also allowed the team to base its operation at Heraklion Air Force Base, which has a runway pointing out to sea in the general direction of Santorini, making it an ideal launching place for the flight attempt.

Once team members had reached Crete, they (like MacCready's cross-channel team before them) began the long wait for a forecast indicating a whole day's good weather between Crete and Santorini. Several pilots were on duty in rotation. On each day that looked promising, one of the pilots prepared to fly and was sitting in the cockpit as the sun rose. If the launch was canceled, a different pilot was on duty the next day. The team waited almost a month before a day arrived when the weather looked good enough to launch.[69]

On April 23, 1988, Kanellos Kanellopoulos was on the pilot schedule for the day. He was a former Greek national champion bicycle racer and already a celebrity in Greece. On that day, with the forecast looking good, all the chase boats were ready, and the weather report from Santorini was favorable. The team decided to launch.

Soon after dawn, Kanellopoulos lifted Daedalus 88 off the runway and headed out over the water. Eventually, he became the first person to fly an HPA out of sight of land. A few miles into the flight, however, his altimeter failed, which as Bryan Allen had previously discovered, makes judging how close one is to the water very difficult. Fortunately for Kanellopoulos, Daedalus was designed to fly much higher over the water than the Gossamer Albatross was, so it was easier for him to use chase boats to judge his height. Three hours and fifty-four minutes (and seventy-four miles) after leaving Crete, Kanellopoulos turned Daedalus to line up with the beach at Santorini. Suddenly a gust of wind shoved him sideways and snapped the tail boom. With the craft now out of control, a wing spar snapped, and Daedalus fell into the water with the fuselage twenty feet from shore. Kanellopoulos emerged unhurt and waded to shore, as team members pulled the wreckage of the craft onto the beach.

As of this writing, Kanellopoulos's flight in Daedalus is the longest distance ever flown by a human-powered airplane.[70] Although the flight ended with a crash, the team considered the flight to be a rousing success. Because they were not burdened by the arbitrary rules of a competition, nothing required their airplane to make a soft landing so many feet beyond the mean high-tide level. Their goal was simply to fly a craft under human power from Crete to the nearest landfall. After flying seventy-four miles over open ocean, the pilot had arrived safely, if unceremoniously, a few steps from the beach. The team had done in real life what until then had only happened in myth: a flight from Crete across the ocean to landfall on muscle power alone.

Ornithopters Move beyond Muscle Power

Early in this chapter, we came to a fork in the road, a parting of the ways between human-powered flight and flapping flying machines, or ornithopters. This split occurred because ornithopter designers understood intuitively what Ursinus and Wilkie would later measure: humans make pitifully weak, heavy airplane engines. After the 1920s, ornithopter builders, like conventional airplane designers, adopted various types of engines in place of human muscle power (except for Alexander Lippisch, who stubbornly held out for muscle power).

Exemplifying the transition from muscle to engine power, Adalbert Schmid built both human- and engine-powered ornithopters in Germany. After constructing models to test his design concepts, he built a person-carrying ornithopter that flew in 1942. It was actually a sort of semi-ornithopter—essentially a sailplane with a small set of flapping wings (with a 10.5-foot wingspan) mounted behind a fixed main wing with a forty-one-foot span. Schmid designed the craft to fly on muscle power, and it made one catapult-launched flight ostensibly under muscle power. Schmid claimed that it maintained level flight for a distance of about 3,000 feet; but an unpowered sailplane might be able to do nearly as well in a decelerating glide from a catapult launch, so I am skeptical that the flapping wings did much more than slightly extend the glide.

Schmid next replaced the pilot's pedaling with a three-horsepower motorbike engine, and he claims the craft took off from the ground and flew for fifteen minutes. After World War II, he modified a conventional sailplane so that the outer parts of the wings flapped, and he claims to have flown this craft under power.[71] Documentation on Schmid's crafts is skimpy, and I must admit to a fair amount of skepticism about his claims. The idea that a three-horsepower engine (less powerful than the average lawnmower engine), working through an inefficient rotary-to-linear transmission and using an inefficient flapping motion, could have produced enough thrust to take off without assistance strains my credulity, although ornithopter enthusiasts accept Schmid's claims at face value.[72]

Between World War II and the 1980s, ornithopters remained firmly in the toy category. A few companies made rubber band–powered flapping-wing toys during this period. The one I remember best was the Tim Bird marketed by G. de Ruymbeke in France and ubiquitous in science museum gift shops in the 1970s and 1980s. Although originally rather expensive for a very light, fragile toy, the Tim Bird, with its birdlike fuselage and adjustable tail, actually climbs under power for a few seconds (until its rubber band motor winds down).

At the dawn of the radio-control model airplane hobby, the engineer who designed the Republic Aviation Seabee seaplane designed and built a model ornithopter. Percival Spencer built his curious engine-powered, R/C biplane ornithopter in 1960 as a proof-of-concept prototype for a person-carrying ornithopter. The upper and lower wings flapped in antiphase: that is, the upper wings flapped down as the lower wings flapped up, and vice versa. Spencer touted this arrangement as a way to avoid oscillations and vibrations from a single pair of flapping wings that would have damaged the delicate radio equipment of the day. His design would have also reduced the bouncing experienced by the pilot of the full-size machine that Spencer envisioned but never built.[73]

Ornithopters Get Bigger . . .

Perhaps the most noteworthy R/C ornithopter was the QN built by Paul MacCready and his company AeroVironment and flown in 1986. QN was a half-scale model of the giant pterosaur *Quetzalcoatlus northropi*, the largest flying animal of all time, and was built to star in an IMAX movie about flight. The AeroVironment team designed QN to look as realistic as possible. It had a rubber-and-Kevlar skin over a carbon fiber skeleton and used its head, with its long beak and crest, as a front rudder for steering. Like most flying animals, QN was unstable, so it was stabilized by a computer using gyroscopic sensors (based on gyroscopes for R/C helicopters, discussed in Chapter 6). The model used electric motors to flap its wings, and the designers tried to emulate the wing shape of real pterosaurs as much as possible.

QN was designed to fly under power by flapping; but in the end, the model was too heavy and the motors too weak to actually maintain level flight by flapping. So the team launched QN with an elastic bungee catapult, like a typical R/C sailplane. Unlike a model sailplane, however, QN has a wingspan of eighteen feet, so it was launched on a cart that stayed on the ground and with an airplane-like stabilizing tail that detached and fell away after launch. (Both QN and its pterosaur namesake were functionally tailless.) Once launched, QN could flap but mostly flew like a sailplane, gaining altitude only in thermals or ridge lift.[74] Ironically, the actual pterosaur, *Quetzalcoatlus*, almost certainly flew the same way because it was far too large to have maintained level flight by muscle-powered flapping.[75] Even more ironically, QN was probably more realistic because it was built and flown in 1986 than it would be if it were built now. At that time, its builders were stuck with brushed DC motors and NiCad batteries. With today's brushless motors and lightweight lithium batteries, an equivalent power system would weigh less than half of the one used in QN. With the standard hobby-grade R/C equipment available today, QN would probably have been able to fly level or even climb by flapping, which would have been most unlike the animal on which it was based.

AeroVironment's QN was quite successful and significant, even though it was a disappointment as an ornithopter. It demonstrated successful soaring using a control mechanism totally unlike any airplane's. It also demonstrated that pterosaurs, like other flying animals, were probably unstable and relied on active reflexes to maintain level flight. Plus, as anyone who has seen the IMAX movie "On the Wing" knows, it was extremely realistic in flight.[76]

And Bigger . . .

After years of research, Jeremy Harris of the Battelle Memorial Institute in Columbus, Ohio and James DeLaurier of the University of Toronto and DeLaurier's students built and flew a large, engine-powered R/C ornithopter in 1991.[77] The Harris-DeLaurier craft had a conventional airplane tail on a rather airplane-like fuselage (fig. 10.8).

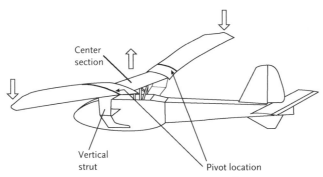

Figure 10.8 Harris and DeLaurier's R/C ornithopter. The wing has a center section that is driven up and down by the motor and two outer panels that pivot like seesaws on vertical struts. When the center section moves up, the wing tips move down.

Its wing at rest looked like a conventional sailplane wing suspended above the fuselage on two struts, although internally it was carefully designed to flex in specific ways. Its flapping mechanism was quite ingenious. The center section of the wing, between the two struts, moved up and down in antiphase to the movement of the wingtips. The wing's angle of attack adjusted itself automatically during the downstroke and the upstroke, and figuring out how to build a wing that passively made those adjustments was one of the reasons that the design required several years of intense research.[78]

The center section of Harris and DeLaurier's wing moved up and down, providing a clever means of converting the engine's rotation into a flapping motion. Moreover, with the center section moving in the opposite direction from the wingtips, the motion helped reduce some of the oscillation of the fuselage caused by the equal and opposite reaction of the wing's up-and-down motion. DeLaurier and Harris intended the R/C ornithopter to be the basis for a full-sized, person-carrying ornithopter. And within four years of the successful R/C ornithopter flight, they built a full-sized ornithopter and soon successfully taxied it under flapping power.

But a series of problems, from control difficulties to trouble finding and keeping test pilots, kept the craft on the ground for about a decade and Harris eventually dropped out of the project. During flight attempts, DeLaurier discovered that, as the craft approached flight speed, the fuselage started bouncing in time with the wing beats. Once this bouncing started, the banging of the landing gear on the runway prevented any further acceleration, so the craft could never quite get all the way up to takeoff speed.[79] Then DeLaurier discovered tiny jet engines for R/C airplanes and realized that the thrust from such an engine might be enough to nudge his ornithopter through its bounces and into the air. On July 8, 2006, with Jack Sanderson at the controls, DeLaurier's ornithopter with jet assistance broke ground and flew down the runway for fourteen seconds at a height of three feet and covering a distance of about 1,000 feet.[80] As far as I know, this is the first time in which a person-carrying ornithopter has risen off the ground under its own power. The use of the jet engine and the short flight time perhaps take a bit of the luster off DeLaurier's achievement. Yet given that the jet engine was probably intended to fly a twenty-pound R/C model rather than a 1,000-pound, full-size airplane, the craft was clearly getting the vast bulk of its thrust from its wings. So DeLaurier is to be congratulated for demonstrating the feasibility (if not the usefulness) of a full-scale ornithopter.

And Then Smaller . . .

In recent years, ornithopter technology has been much more successful at the small end of the size spectrum. For example, in the late 1990s, the first R/C ornithopter kits came onto the hobby market. One of the first was Sean Kinkaide's simple, open-framework Skybird, powered by a small R/C engine. Soon after, several electric-powered kits became available.

At about the same time, a number of research groups at universities and in industry became interested in designing very small R/C or autonomous flying vehicles. Their ultimate goal was to produce a tiny, autonomous surveillance and reconnaissance craft. The Defense Advanced Research Projects Agency (DARPA) began funding research into microrobot flyers, aiming for a flying vehicle with at most a six-inch wingspan that could carry a camera and had an endurance of up to two hours.

A team drawn from the California Institute of Technology and AeroVironment developed the Microbat, a tiny R/C ornithopter with a six-inch wingspan powered by an electric motor. They named it after a bat because its wings are thin sheets of plastic film supported by thin composite rods, similar to the way in which a bat's wing membrane is supported by its finger bones. Although the Microbat has a conventional, airplane-like tail, its rudder-control mechanism is eerily biological. It is connected to two lengths of "muscle wire" made of a metal alloy that shortens when a voltage is applied. These two wires move the rudder in exactly the same way that a pair of antagonistic muscles bends and straightens your knee.[81]

A number of tiny autonomous flyers came out of the DARPA program, but none could carry a camera or fly for anywhere near two hours.[82] Although DARPA has apparently given up (for now) on the idea of robotic aerial surveillance ornithopters small enough to be mistaken for birds or insects, a few research groups continue to pursue the goal of tiny, insect-like flying machines. For example, Robert Michelson and his students at the Georgia Institute of Technology have been working for several years on his Entomopter. Somewhat reminiscent of a dragonfly, the Entomopter concept uses two wings, a front one and a back one. Rather than being hinged in the middle at the wing root, each wing has a rigid leading edge all the way from tip to tip, and it "flaps" by rocking from side to side on the fuselage like a seesaw. The front and hindwings rock in antiphase, so any torque on the fuselage caused by one wing tilting to one side is canceled by the torque of the other wing rocking to the other side.

Taking his simulation of nature even further, Michelson has spent several years developing a "chemical reciprocating muscle" to power the Entomopter. This is a linear actuating device (like a muscle) that runs on chemical fuel (like a muscle) without actually burning it (like a muscle) and is inherently much better suited to move flapping wings than an electric motor is. Although he has flown scaled-up, rubber band–powered Entomopter models to validate the flapping mechanism, his prototype chemical muscles are too big so far to power the two-ounce vehicle he has designed. As of this writing, the Entomopter has not yet flown in its definitive form.[83]

Ron Fearing and his students and colleagues at the University of California at Berkeley are aiming even smaller. Drawing on recent research on house fly and fruit fly aerodynamics, they are designing a micromechanical flying insect (MFI), a tiny flying machine with a one-inch wingspan. Its wings are driven by piezo crystals (like the ones in solid-state gyroscopes, discussed in Chapter 6) mounted on a carbon fiber frame and transmission. The wings are rather insect-like, with polyester "veins" and a polyester membrane with a thickness that is roughly one-fiftieth the diameter of a human hair. So far, they have built single-wing prototypes to measure lift and thrust production, and they are currently in the process of developing smaller and lighter frameworks and actuators as well as lighter wings.[84]

In a sense, we have come full circle. First, people tried to fly like birds by flapping feather-covered arms or using machines with wings that the pilot flapped with his own muscle power. Then, with the realization that lift and thrust need not be produced by the same device, aviation diverged from flying animals. Even muscle-powered flying machines used propellers and fixed wings. Now, as researchers and inventors strive to develop ever-smaller flying machines, they have gone back to emulating nature. The smallest flying machines, micro air vehicles (MAVs), are not all ornithopters by any means. Even AeroVironment has produced at least two propeller-driven MAVs in addition to the Microbat. As MAVs get smaller, however, combining lift and thrust production with flapping wings begins to look more attractive. As a sign of growing interest in the concept, the International Micro Air Vehicle competition, an annual student event at the University of Florida, added an ornithopter category in 2004.

The smaller that MAVs get, the more often their designers look to nature for design inspiration: bat wings for the Microbat, muscles for the Entomopter, fly aerodynamics and wing movements for the MFI. Just as with animals, as flying machines get really tiny, their inherently low wingloading and high power for their weight make flapping wings practical and lessons from Mother Nature relevant. A decade ago, I would have argued that almost no aviation technology has been copied from nature. Now, in the MAV field, the reverse is true. The smaller and more advanced the vehicle, the more its designers borrow from flying animals.

In memoriam: Dr. Paul B. MacCready—champion model builder and sailplane pilot; builder of human-powered aircraft, solar-powered aircraft, a flying pterosaur model, numerous unmanned air vehicles, and micro air vehicles; and winner of multiple Kremer prizes—died on August 28, 2007, at the age of eighty-one.

Epilogue
So Why Don't Jumbo Jets
Flap Their Wings?

A nything that flies, whether animal or machine, obeys the same basic aerodynamic principles (see Chapter 2). When a cambered wing moves through the air at a positive angle of attack, the air does not care whether the wing is made of feathers or aluminum: the air flow still produces lift. No matter how big or small the wing, if you move it faster or tilt it to a higher angle of attack, then the wing generates more lift. True, size does matter: as wings get bigger, they become more efficient; and as wings get smaller, they get disproportionately lighter. But these size effects are quantitative, not qualitative. A wing works more or less the same way, whether attached to a sparrow or a jet airliner.

Effective wings are the essential feature of any flyer. Wings may well be the only significant feature shared by a sparrow and a Spitfire or by a bumblebee and a Boeing 747. Moreover, all birds get by with a single pair of very versatile wings, whereas different flying machines employ a diversity of specialized wings—propeller blades, helicopter rotors, even compressor and turbine blades in jet engines—in addition to the obvious fixed wings of most airplanes.

Flying animals and aircraft also share a history of change over time, from more primitive designs that could barely get aloft to present-day designs that routinely fly enormous distances. The agent of change in flying animals is evolution by natural selection. If being a better flyer is an advantage, animals that are better flyers will tend to reproduce more successfully, and eventually the whole population will become better flyers. In aviation, designers usually set out to design an aircraft to perform a particular task as efficiently as possible. Sometimes that means being first at something—thermal soaring, for example, or hovering, or flying across the Atlantic—but usually it means performing the task better than previous designs have. So both technological progress and natural selection tend to lead to more accomplished flyers. In nature and in aviation, flyers need to perform as well or better than their competitors, or their competitors will replace them.

In spite of these similarities, the differences between flying animals and human-designed aircraft are vast. If flyers are all constrained by the same physics, what explains these differences?

Though airplanes and animals have both evolved from less capable to more capable designs, the mechanisms and timing of those changes are radically different. As with any technology since the beginning of the scientific revolution, airplane designers have consciously designed airplanes to perform particular tasks, such as to soar, to train pilots, to carry cargo, and so on. Airplane designs improve as engineers figure out new and better ways to build structures or perform functions. Technological change is thus an intentional, goal-directed process. The designer aims toward a particular goal and uses human intelligence to anticipate both drawbacks and advantages and to make trade-offs to achieve a useful machine. When a designer sets out to build, say, a new training airplane, he or she will most likely end up with some sort of trainer and not a helicopter or an airliner or a bomber.

Birds, bats, and insects are products of biological evolution, which is much more constrained than technological evolution is. If an airplane designer wants to make an airplane part out of fiberglass instead of aluminum, nothing fundamental prevents the change. In contrast, a bat might benefit from having a skeleton made of graphite-epoxy composite instead of bone, but such a change is impossible: the bat, being a mammal, is constrained by its ancestry to have a skeleton made of bone. Similarly, insect wings are modified exoskeleton; and because insect exoskeletons are made of protein and chitin, insects have no way to build wings out of aluminum or plastic. As long as insects have wings, they will be made out of variations on the material that forms their exoskeleton.

In spite of these constraints, animals do change over time, but the changes are not goal-directed. Neither, however, are these changes random. True, random mutations provide grist for the mill of evolution, but natural selection is anything but random. Most mutations are either neutral (have no effect) or harmful, and harmful ones will quickly die out. Beneficial mutations are exceedingly rare. If a beneficial mutation makes an animal a little bit better at doing something, such as flying for longer distances, and if longer flights are an advantage, then that ability should spread through the population until all the animals in the population can fly farther. The key point here is that the animals are more successful with this new ability. Natural selection tends to produce animals that are very well adapted for a particular set of conditions rather than to make changes to achieve a specific goal or task. So if improved flight agility is an advantage in some conditions, flying animals in those conditions will probably evolve to be more maneuverable. In some conditions, superior flight ability is an advantage; so house flies, swifts, and greater noctule bats have evolved. In other conditions, flight is actually detrimental; so ostriches and fleas have evolved flightlessness, even though their ancestors flew.

Flying animals, unlike airplanes, have been around long enough to experience fundamental changes in conditions on Earth. About 280 million years ago, a large increase in the proportion of oxygen so improved the effectiveness of diffusion-limited insect respiratory systems that gigantic insects evolved—mayflies with eight-inch wingspans and dragonflies with twenty-eight-inch wingspans. Some scientists speculate that this atmospheric change, by affecting the air's density and viscosity, may even have made flight easier and thus stimulated the early evolution of flight in insects.[1] After nearly 100 million years, the oxygen level fell back to lower values, too low to support giant insects, which went extinct. Small insects had become such accomplished flyers during the period of high oxygen, however, that they prospered even at lower oxygen levels.

So airplanes and animals both evolve, but the timescales of biological and technological change are vastly, incredibly different. Charles Darwin conceived of evolution as gradual and continuous, but we now know that natural selection often seems to work in fits and starts, with no obvious changes for long periods interspersed with shorter periods of rapid change.[2] *Short* and *rapid* in geological terms are still tremendously slow. Whereas long periods of little change can last tens or even hundreds of millions of years (dragonflies have changed very little in more than 200 million years), periods of faster change still last from tens of thousands to millions of years. In other words, a period of "rapid" change of only 20,000 years, on the short end of the scale, is still several times longer than all of recorded human history. At the other end of the scale, such rapid-change episodes can last for millions of years, longer than the entire existence of modern humans as a species. For example, the best fossil evidence to date suggests that beaked, modern-looking birds took tens of millions of years to evolve from rather reptilian, ancestral, toothed birds such as *Archaeopteyx*.

In contrast, the scientific revolution began barely four hundred years ago. In that time, the pace of technological change has been so much faster than evolutionary biological change that the difference in timing is nearly inconceivable. Whereas insects took more than 100 million years to evolve from ancestral, wingless forms to house flies, engineers needed only about fifteen years to take airplanes from the Wright Flyer to machines that could loop and roll, or carry several passengers, or fly hundreds of miles. Airplane designers needed only twenty-five years, barely more than one human generation, to progress from wooden, wire-braced, hundred-mile-per-hour biplanes such as Sopwith Camels to all-metal airplanes with top speeds of more than four hundred miles per hour, more than half the speed of sound. The first jets also appeared at the end of that twenty-five-year period, flying even faster. Aviation took fewer than five decades to go from the Wright brothers' first flight to the first flight that was faster than the speed of sound. Within seventy years of the Wrights' flight—a single human lifetime—aviation progressed from a craft that could carry only the pilot and barely fly a half-mile to the first Boeing 747 jumbo jet that could carry hundreds of people across the Atlantic Ocean. This enormously rapid progress is due to the goal-directed nature of human technology as well as the predictive ability and creativity of human intelligence. Rather than waiting for the right combination of appropriate genetic mutations and fortuitous selective conditions, a human designer can perceive a need, envision a new solution, and design and build an airplane in a few years or less—much less than one human lifetime, let alone a generation.

Probably the single most important factor enforcing differences between nature's flyers and flying machines is their respective power sources (see Chapter 3). Animals use muscle, so they cannot use propellers; but flapping their wings is a natural, adaptable movement for animals. Airplanes use rotating engines, making fixed wings and a separate thrust producer a logical solution. (In this context, human-powered airplanes count as machines due to their rotating drive shafts, even though they are ultimately muscle-powered.) This difference in power sources, indeed, is the main answer to the question in the title of this epilogue.

As we saw in Chapter 10, some flying machines do actually flap their wings, although few ornithopters have actually flown successfully. If we discount various full-sized machines that flew mainly as gliders or with catapult launches and accord DeLaurier's machine the honor of the first person-carrying ornithopter to have unequivocally risen off the ground on flapping wings, consider the timing. DeLaurier's flight was more than a century after the Wright brothers' first flight and nearly forty years after the first person landed on the moon.

This timing suggests a couple of different possibilities. Perhaps flapping flight is simply much more complex and difficult than flight with fixed wings. Birds make it look easy, but after all, they have had more than 150 million years to get good at flapping flight. Alternatively, ornithopters may be machines without a mission. If they have little or no practical utility, then serious engineers and designers may not want to waste effort on trying to build machines that are not an improvement on any conventional aircraft. In my opinion, both possibilities are more or less true. Flapping flight is more complex and difficult than most ornithopter designers have realized until recently. Even so, success would have come sooner if ornithopters had any practical applications. DeLaurier and other ornithopter enthusiasts contend that the craft might fly more quietly than conventional airplanes, but so far that is an untested hypothesis. As far as I know, ornithopters do not have any abilities that improve on anything done by conventional aircraft.

In contrast, when flying machines get very small, flapping can actually become an advantage. When flying machines shrink to the size of pigeons or even smaller, tiny propellers are so inefficient that flapping wings may better them in efficiency. Tiny propellers must accelerate a small volume of air by a large amount, which is inefficient compared with flapping wings, which can accelerate a large volume of air by a small amount. Moreover, small propellers must spin exceedingly fast, which leads to the problem of supersonic tips and huge power requirements. In contrast, the small wings of a bird-sized ornithopter are easier to flap than they are on larger machines because these tiny wings can be very light yet still strong enough to withstand aerodynamic stresses. Flapping wings combine lift and thrust production into one device, which might potentially simplify the design of extremely small flying machines. In the realm of micro aerial vehicles, the advantage of flapping appears to be great enough to allow ornithopter designs to compete successfully with, or even out-compete, more conventional designs.

In the end, what truly sets birds apart from airplanes is versatility versus efficiency. Engineers design airplanes to carry out particular tasks, so airplanes tend to be quite specialized. State-of-the-art sailplanes can have stunningly high lift-to-drag ratios—twice as high as anything in nature—because all a sailplane really needs to do is soar. Similarly, a Boeing 747 can haul huge loads of passengers over enormous distances, but that is basically all it can do. It cannot perform aerobatics, nor soar in thermals, nor drive on the ground on or off roads.

Flying animals trade great efficiency for versatility. No bird has as high a lift-to-drag ratio as a modern sailplane does, nor can any bird carry dozens of other birds on flights from one place to another. But all birds can move around on the ground reasonably well, many can hover, and all are far more maneuverable than any airplane. Birds, bats, and insects do not have the luxury of operating only from airports; they need to be able to take off and land almost anywhere, from tree branches to cave ceilings to seaside cliffs. Soaring birds still need to be able to flap, and even dyed-in-the-wool flappers such as swallows and dragonflies need to be able to glide. In short, flying animals need to be able to do many things fairly well, in contrast to airplanes, which can often do one or two things extremely well but very little else. Airplanes can be quite successful even when they are only good for one particular type of flying, but animals cannot afford to be so specialized. Consider ducks, which fly, walk, and swim underwater, or terns, which hover, soar, and swim. Finally, the most important difference in versatility is that all animals can reproduce, whereas self-replicating machines are still the stuff of science fiction.

The old adage is "jack of all trades, master of none," but thinking of birds as "jacks" in comparison with "masterful" aircraft is misleading. Airplanes can, indeed, be masters at a given task, but flying animals can also be near-masters at particular types of flight. Hummingbirds and honeybees hover masterfully, vultures and eagles are expert at soaring, and insect-hawking bats are mightily maneuverable. But even less-specialized flyers such as robins and June beetles do so many different tasks reasonably well, including various forms of flight, that their versatility itself is a form of mastery.

To fundamentally terrestrial beings such as ourselves, flight is a wonderful proposition. It literally opens up a new, third dimension to locomotion. Yet flight is not a trivial process. If flight were easy, most animals would fly, and the ancient Romans or Chinese would have built airplanes. Although very exacting, the aerodynamic principles that make up the rules of the game do allow a wide range of design possibilities. Consider the Boeing 737 airliner and the bumblebee with which I began the first chapter of this book. As different as they are, both are very successful, and both owe that success to their ability to fly.

NOTES

Chapter 1: Flying Animals and Flying Machines

1. John J. Bertin and Michael L. Smith, *Aerodynamics for Engineers* (Englewood Cliffs, N.J.: Prentice Hall, 1979), 37–163.

2. Jeremy M. V. Rayner, "Form and Function of Avian Flight," *Current Ornithology* 5 (1988): 1–66.

3. Paul Amos Moody, *Introduction to Evolution*, 3d ed. (New York: Harper and Row, 1970), 11–20.

4. Stephen J. Gould and Niles Eldredge, "Punctuated Equilibria: The Tempo and Mode of Evolution Reconsidered," *Paleobiology* 3, no. 2 (1977): 115–51.

5. David E. Alexander, *Nature's Flyers: Birds, Insects, and the Biomechanics of Flight* (Baltimore: Johns Hopkins University Press, 2002), 167–68.

6. Pat Shipman, *Taking Wing: Archaeopteryx and the Evolution of Bird Flight* (New York: Simon and Schuster, 1998).

7. John H. Ostrom, "Bird Flight: How Did It Begin?," *American Scientist* 67 (1979): 46–56.

8. Alan Feduccia, *The Origin and Evolution of Birds*, 2d ed. (New Haven, Conn.: Yale University Press, 1999), 110–12; Jeremy M. V. Rayner, "The Evolution of Vertebrate Flight," *Biological Journal of the Linnean Society* 34 (1988): 269–87; U. M. Norberg, *Vertebrate Flight: Mechanics, Physiology, Morphology, Ecology and Evolution*, (New York: Springer-Verlag, 1990), 257–59, 265–66.

9. James P. Garner, Graham K. Taylor, and Adrian L. R. Thomas, "On the Origins of Birds: The Sequence of Character Acquisition in the Evolution of Avian Flight," *Proceedings of the Royal Society of London B* 266 (1999): 1259–66.

10. Keith Scholey, "The Evolution of Flight in Bats," in *Bat Flight—Fledermausflug*, ed. W. Nachtigall (Stuttgart, Germany: Fischer, 1986), 1–12.

11. Charles H. Gibbs-Smith, *The Aeroplane: An Historical Survey of Its Origins and Development* (London: Her Majesty's Stationery Office, 1960), 23–54.

12. Charles H. Gibbs-Smith, *Aviation: An Historical Survey from Its Origins to the End of World War II* (London: Her Majesty's Stationery Office, 1970), 105–37; John David Anderson, *The Airplane: A History of Its Technology* (Reston, Va.: American Institute of Aeronautics and Astronautics, 2002).

13. Gibbs-Smith, *Aviation*, 139–50.

14. Peter L. Jakab, *Visions of a Flying Machine: The Wright Brothers and the Process of Invention*, ed. D. A. Pisano (Washington, D.C.: Smithsonian Institution Press, 1990), 6–7.

15. Tom D. Crouch, *The Bishop's Boys: A Life of Wilbur and Orville Wright* (New York: Norton, 1989), 94–115.

16. John David Anderson, *Inventing Flight: The Wright Brothers and Their Predecessors* (Baltimore: Johns Hopkins University Press, 2004), 133–35.
17. Crouch, *The Bishop's Boys*, 244–45.
18. Anderson, *The Airplane*, 121.

Chapter 2: Hey, Buddy, Need a Lift?

1. Jeremy M. V. Rayner, "The Mechanics of Flapping Flight in Bats," in *Recent Advances in the Study of Bats*, ed. M. B. Fenton, P. Racey, and J. M. V. Rayner (Cambridge, U.K.: Cambridge University Press, 1987), 23–42.
2. John J. Bertin and Michael L. Smith, *Aerodynamics for Engineers* (Englewood Cliffs, N.J.: Prentice Hall, 1979), 180.
3. Peter B. S. Lissaman, "Low-Reynolds-Number Airfoils," *Annual Review of Fluid Mechanics* 15 (1983): 223–39, 225.
4. S. Ward, U. Moeller, J. M. V. Rayner, D. M. Jackson, D. Bilo, W. Nachtigall, and J. R. Speakman, "Metabolic Power, Mechanical Power and Efficiency during Wind Tunnel Flight by the European Starling *Sturnus vulgaris*," *Journal of Experimental Biology* 204, no. 19 (2001): 3311–22; Steven Vogel, *Life in Moving Fluids: The Physical Biology of Flow*, 2d ed. (Princeton, N.J.: Princeton University Press, 1994), 249.
5. Henk Tennekes, *The Simple Science of Flight: From Insects to Jumbo Jets* (Cambridge, Mass.: MIT Press, 1996), 70, 83.
6. Tennekes, *Simple Science*, 83; Vogel, *Moving Fluids*, 249.
7. David E. Alexander, *Nature's Flyers: Birds, Insects, and the Biomechanics of Flight* (Baltimore: Johns Hopkins University Press, 2002), 40.
8. Ira H. A. Abbott and Albert Edward von Doenhoff, *Theory of Wing Sections, Including a Summary of Airfoil Data*, reprint, corrected ed. (New York: Dover, 1959), 3.
9. Bertin and Smith, *Aerodynamics*, 100.
10. John David Anderson, *The Airplane: A History of Its Technology* (Reston, Va.: American Institute of Aeronautics and Astronautics, 2002), 123.
11. Tennekes, *Simple Science*, appendix.
12. Vance A. Tucker, "Drag Reduction by Wing Tip Slots in a Gliding Harris' Hawk, *Parabuteo unicinctus*," *Journal of Experimental Biology* 198 (1995): 775–81; Vance A. Tucker, "Gliding Birds: Reduction of Induced Drag by Wing Tip Slots between the Primary Feathers," *Journal of Experimental Biology* 180 (1993): 285–310.
13. Jan Roskam and C. Edward Lan, *Airplane Aerodynamics and Performance* (Lawrence, Kans.: DAR, 1997), 56.
14. Philip C. Withers, "An Aerodynamic Analysis of Bird Wings As Fixed Aerofoils," *Journal of Experimental Biology* 90, no. 1 (1981): 143–62; Steven Vogel, "Flight in *Drosophila*. III. Aerodynamic Characteristics of Fly Wings and Wing Models," *Journal of Experimental Biology* 46 (1967): 431–43.
15. *U.S. Code of Federal Regulations*, title 14, parts 23 and 25 (http://ecfr.gpoaccess.gov), accessed August 30, 2007.
16. Ibid.
17. John D. Currey, *The Mechanical Adaptations of Bones* (Princeton, N.J.: Princeton University Press, 1984), 237; Andrew A. Biewener and Kenneth P. Dial, "In Vivo Strain in the Humerus of Pigeons (*Columba livia*) during Flight," *Journal of Morphology* 225 (1995): 61–75; Sean J. Kirkpatrick, "Scale Effects on the Stresses and Safety Factors in the Wing Bones of Birds and Bats," *Journal of Experimental Biology* 190, no. 1 (1994): 195–215.
18. Anderson, *The Airplane*, 161–70.
19. Ibid., 260–62.

20. Robert M. Jones, *Mechanics of Composite Materials*, 2d ed. (Philadelphia: Taylor and Francis, 1999), 3.
21. Ibid., 15–25.
22. Theodore von Kármán, *Aerodynamics* (New York: McGraw-Hill, 1963), 41–43.
23. Sighard F. Hoerner and Henry V. Borst, *Fluid-Dynamic Lift* (Bricktown, N.J.: Hoerner Fluid Dynamics, 1975), 4.12.
24. Ibid., 2.1–2.4.
25. Bertin and Smith, *Aerodynamics*, 21; Vogel, *Life in Moving Fluids*, 84–87.
26. Lissaman, "Low-Reynolds-Number Airfoils."
27. Ibid.
28. Tennekes, *Simple Science.*
29. Charles H. Gibbs-Smith, *Aviation: An Historical Survey from Its Origins to the End of World War II* (London: Her Majesty's Stationery Office, 1970), 182.
30. Anderson, *The Airplane*, 171–81, 201–4.
31. Jones, *Mechanics of Composite Materials*, 34–45.
32. Don Downie and Julia Downie, *The Complete Guide to Rutan Aircraft*, 3d ed. (Blue Ridge Summit, Penn.: Tab Books, 1987), 89–95.
33. Andrei K. Brodsky, *The Evolution of Insect Flight* (Oxford: Oxford University Press, 1994), 16–25.
34. Karl Herzog, *Anatomie und Flugbiologie der Vögel* (Stuttgart, Germany: Fischer, 1968), 55.
35. Robert J. Raikow, "Locomotor Systems," in *Form and Function in Birds*, ed. A. S. King and J. McLelland (New York: Academic Press, 1985), 57–148.
36. R. Åke Norberg, "Function of Vane Asymmetry and Shaft Curvature in Bird Flight Feathers; Inferences on Flight Ability of *Archaeopteryx*," in *The Beginnings of Birds*, ed. M. K. Heckt, J. H. Ostrom, G. Viohl, and P. Wellnhofer, (Eichstätt, Germany: Freunde des Jura-Museums Eichstätt, 1985), 303–18.
37. Alan Feduccia, *The Origin and Evolution of Birds*, 2d ed. (New Haven, Conn.: Yale University Press, 1999), 112.
38. Alfred J. Ward-Smith, *Biophysical Aerodynamics and the Natural Environment* (New York: Wiley, 1984), 143–44.
39. S. M. Swartz, M. S. Groves, H. D. Kim, and W. R. Walsh, "Mechanical Properties of Bat Wing Membrane Skin," *Journal of Zoology* [London] 239, no. 2 (1996): 357–78.
40. Thomas H. Kunz and M. Brock Fenton, *Bat Ecology* (Chicago: University of Chicago Press, 2003), 307–10.
41. Wann Langston, Jr., "Pterosaurs," *Scientific American* 224 (1981): 122–36.
42. Kevin Padian and Jeremy M. V. Rayner, "The Wings of Pterosaurs," *American Journal of Science* 293A (1993): 91–166.
43. D. W. Whitman, L. Orsak, and E. Greene, "Spider Mimicry in Fruit Flies (Diptera: Tephritidae): Further Experiments on the Deterrence of Jumping Spiders (Araneae: Salticidae) by *Zonosemata vittigera* (Coquillett)," *Annals of the Entomological Society of America* 81, no. 3 (1988): 532–36.
44. Annelli Hoikkala and Philip Welbergen, "Signals and Responses of Females and Males in Successful and Unsuccessful Courtships of Three Hawaiian Lek-Mating *Drosophila* Species," *Animal Behaviour* 50, no. 1 (1995): 177–90.
45. Richard O. Prum, "Sexual Selection and the Evolution of Mechanical Sound Production in Manakins (Aves: Pipridae)," *Animal Behaviour* 55, no. 4 (1998): 977–94.
46. A. K. Eggert and S. K. Sakaluk, "Sexual Cannibalism and Its Relation to Male Mating Success in Sagebrush Crickets, *Cyphoderris strepitans* (Haglidae, Orthoptera)," *Animal Behaviour* 47, no. 5 (1994): 1171–77.

47. Daniel K. Riskin, S. Parsons, W. A. Schutt, Jr., G. G. Carter, and J. W. Hermanson, "Terrestrial Locomotion of the New Zealand Short-Tailed Bat *Mystacina tuberculata* and the Common Vampire Bat *Desmodus rotundus*," *Journal of Experimental Biology* 209, no. 9 (2006): 1725–36.

Chapter 3: Power

1. Merran Williams, "The 156-Tonne Gimli Glider," *Flight Safety Australia* 7, no. 3 (2003): 22–27.

2. Steven Vogel, *Prime Mover: A Natural History of Muscle* (New York: Norton, 2001), 4, 61.

3. Ibid., 96.

4. Matthew W. Bundle, Kacia S. Hansen, and Kenneth P. Dial, "Does the Metabolic Rate-Flight Speed Relationship Vary among Geometrically Similar Birds of Different Mass?" *Journal of Experimental Biology* 210, no. 6 (2007): 1075–83; K. P. Dial, A. A. Biewener, B. W. Tobalske, and D. R. Warrick, "Mechanical Power Output of Bird Flight," *Nature* 390, no. 6,655 (1997): 67–70; S. Ward, U. Moeller, J. M. V. Rayner, D. M. Jackson, D. Bilo, W. Nachtigall, and J. R. Speakman, "Metabolic Power, Mechanical Power and Efficiency during Wind Tunnel Flight by the European Starling *Sturnus vulgaris*," *Journal of Experimental Biology* 204, no. 19 (2001): 3311–22.

5. Peng Chai, Johnny S. C. Chen, and Robert Dudley, "Transient Hovering Performance of Hummingbirds under Conditions of Maximal Loading," *Journal of Experimental Biology* 200 (1997): 921–29; Robert Dudley and Charles P. Ellington, "Mechanics of Forward Flight in Bumblebees. II. Quasisteady Lift and Power Requirements," *Journal of Experimental Biology* 148 (1990): 53–88; Jeremy M. V. Rayner, "Estimating Power Curves of Flying Vertebrates," *Journal of Experimental Biology* 202, no. 23 (1999): 3449–61; Vance A. Tucker, "Respiratory Exchange and Evaporative Water Loss in the Flying Budgerigar," *Journal of Experimental Biology* 48 (1968): 67–87.

6. Jan Roskam and C. Edward Lan, *Airplane Aerodynamics and Performance* (Lawrence, Kans.: DAR, 1997), 201–2.

7. John David Anderson, *The Airplane: A History of Its Technology* (Reston, Va: American Institute of Aeronautics and Astronautics, 2002), 285–88.

8. Roskam and Lan, *Airplane Performance*, 232.

9. "De Havilland Canada D.H.C.2 'Beaver I'" (http://www.aero-web.org/specs/dehacana/dhc2i.htm), accessed April 16, 2008; Smithsonian National Air and Space Museum, "Piper J-3 Cub" (http://collections.nasm.si.edu/code/emuseum.asp?profile=objects&newstyle=single&quicksearch=A19771128000), accessed April 16, 2008.

10. Jane's Information Service, *Jane's All the World's Aircraft* (http://jawa.janes.com/docs/jawa/search.jsp), accessed March 31, 2008.

11. Steven Vogel, *Life's Devices: The Physical World of Animals and Plants* (Princeton, N.J.: Princeton University Press, 1988), 274–76.

12. John D. DeLaurier and J. M. Harris, "A Study of Mechanical Flapping-Wing Flight," *Aeronautical Journal* 97 (1993): 277–86.

13. Brendan J. Borrell and Matthew J. Medeiros, "Thermal Stability and Muscle Efficiency in Hovering Orchid Bees (Apidae: Euglossini)," *Journal of Experimental Biology* 207, no. 17 (2004): 2925–33; Ward et al., "Metabolic Power of the Starling," 3311–22.

14. Werner Nachtigall, *Insects in Flight: A Glimpse behind the Scenes in Biophysical Research* (New York: McGraw-Hill, 1974); U. M. Norberg, "Aerodynamics of Hovering Flight in the Long-Eared Bat *Plecotus auritus*," *Journal of Experimental Biology* 65, no. 2 (1976): 459–70; H. D. J. N. Aldridge, "Body Accelerations during the Wingbeat in Six Bat

Species: The Function of the Upstroke in Thrust Generation," *Journal of Experimental Biology* 130 (1987): 275–94.
15. Anthony S. King and John McLelland, *Birds: Their Structure and Function*, 2d ed. (Philadelphia: Bailliáere Tindall, 1984), 60–63.
16. David E. Alexander, *Nature's Flyers: Birds, Insects, and the Biomechanics of Flight* (Baltimore: Johns Hopkins University Press, 2002), 78–85.
17. Anderson, *The Airplane*, 245–50.
18. Bill Gunston, *World Encyclopaedia of Aero Engines: All Major Aircraft Power Plants, from the Wright Brothers to the Present Day*, 4th ed. (Newbury Park, Calif.: Haynes North America, 1998), 131–32.
19. Ibid., 111.
20. Anderson, *The Airplane*, 285–88.
21. Roskam and Lan, *Airplane Aerodynamics*, 233–35.
22. Anderson, *The Airplane*, 53–56, 72–74.
23. Leonard Bridgman, *Jane's All the World's Aircraft, 1945–46*, vol. 34 (New York: Macmillan, 1946), 54D–55D.
24. C. D. B. Bryan, *The National Air and Space Museum*, 2d ed. (New York: Abrams, 1988), 69–76.
25. John J. Bertin and Michael L. Smith, *Aerodynamics for Engineers* (Englewood Cliffs, N.J.: Prentice Hall, 1979), 377–88.
26. M. Milstein, "Superduperjumbo," *Air and Space Smithsonian* 21, no. 2 (2005): 22–27.

Chapter 4: To Turn or Not to Turn

1. David E. Alexander, *Nature's Flyers: Birds, Insects, and the Biomechanics of Flight* (Baltimore: Johns Hopkins University Press, 2002), 121–23.
2. Jan Roskam and C. Edward Lan, *Airplane Aerodynamics and Performance* (Lawrence, Kans.: DAR, 1997), 606–7.
3. Thomas Charles Gillmer and Bruce Johnson, *Introduction to Naval Architecture* (Annapolis, Md.: Naval Institute Press, 1982), 281–83.
4. Alexander, *Nature's Flyers*, 121–27.
5. David E. Alexander, "Wind Tunnel Studies of Turns by Flying Dragonflies," *Journal of Experimental Biology* 122 (1986): 81–98.
6. Peter L. Jakab, *Visions of a Flying Machine: The Wright Brothers and the Process of Invention*, ed. D. A. Pisano (Washington, D.C.: Smithsonian Institution Press, 1990), 28–29, 49; Tom D. Crouch, *The Bishop's Boys: A Life of Wilbur and Orville Wright* (New York: Norton, 1989), 168–69.
7. John D. Anderson, *The Airplane: A History of Its Technology* (Reston, Va: American Institute of Aeronautics and Astronautics, 2002), 127.
8. Jakab, *Flying Machine*, 42–46.
9. Anderson, *The Airplane*, 120–22.
10. Jakab, *Flying Machine*, 51–58.
11. Crouch, *The Bishop's Boys*, 169–70.
12. Anderson, *The Airplane*, 102–4.
13. Jakab, *Flying Machine*, 172–73.
14. Crouch, *The Bishop's Boys*, 238–39.
15. Ibid., 240.
16. Tom Crouch, "The Thrill of Invention," *Air and Space Smithsonian* 13, no. 1 (1998): 22–30.
17. Charles H. Gibbs-Smith, *The Aeroplane: An Historical Survey of Its Origins and Development* (London: Her Majesty's Stationery Office, 1960), 184.

18. Ibid., 181–83.
19. Ibid., 184.
20. Roskam and Lan, *Airplane Aerodynamics*, 132.
21. Jan Roskam, *Airplane Flight Dynamics and Automatic Flight Controls*, vol. 1 (Ottawa, Kans.: Roskam Aviation and Engineering, 1979), 140.
22. Martin Simons, *Model Aircraft Aerodynamics* (Hemel Hempstead, Hertfordshire, U.K.: Argus, 1987), 179.
23. Greg Goebel, "Unmanned Aerial Vehicles" (http://www.vectorsite.net/twuav_01. html#m1), accessed April 30, 2008.
24. Simons, *Model Aircraft Aerodynamics*, 177–79, 192.
25. F. E. C. Culick and Henry R. Jex, "Aerodynamics, Stability and Control of the 1903 Wright Flyer," in *The Wright Flyer: An Engineering Perspective*, ed. H. S. Wolko (Washington, D.C.: Smithonian Institution Press, 1987), 19–43.
26. Simons, *Model Aircraft Aerodynamics*, 154; Karl Nickel and Michael Wohlfahrt, *Tailless Aircraft in Theory and Practice*, AIAA Education Series (Washington, D.C.: American Institute of Aeronautics and Astronautics, 1994), 85.
27. Roskam, *Airplane Flight Dynamics*, 140–41.
28. John Maynard Smith, "The Importance of the Nervous System in the Evolution of Animal Flight," *Evolution* 6 (1952): 127–29.
29. Culick and Jex, "Stability of the Wright Flyer," 21–22, 30–31.
30. Anderson, *The Airplane*, 356.
31. Malcolm J. Abzug and Eugene E. Larrabee, *Airplane Stability and Control: A History of the Technologies That Made Aviation Possible*, 2d ed., Cambridge Aerospace Series (Cambridge, U.K.: Cambridge University Press, 2002), 313–14.

Chapter 5: A Tale of Two Tails

1. Karl Herzog, *Anatomie und Flugbiologie der Vögel* (Stuttgart, Germany: Fischer, 1968), 87.
2. L. M. Chiappe, J. Shu'An, J. Qiang, and M. A. Norell, "Anatomy and Systematics of the Confuciusornithidae (Theropoda: Aves) from the Late Mesozoic of Northeastern China," *Bulletin of the American Museum of Natural History* 242 (1999): 1–86.
3. Paul D. Kyle and Georgean Z. Kyle, *Chimney Swifts: America's Mysterious Birds above the Fireplace*, Louise Lindsey Merrick Natural Environment Series, no. 37 (College Station: Texas A&M University Press, 2005), 15.
4. Malcolm J. Abzug and Eugene E. Larrabee, *Airplane Stability and Control: A History of the Technologies That Made Aviation Possible*, 2d ed., Cambridge Aerospace Series (Cambridge, U.K.: Cambridge University Press, 2002), 5.
5. Ibid.
6. Ibid.
7. William K. Kershner, *The Student Pilot's Flight Manual: From First Flight to the Private Certificate*, 5th ed. (Ames: Iowa State University Press, 1979), 44–45.
8. Ibid., 94–95.
9. *Private Pilot Manual* (Denver: Jeppesen-Sanderson, 1981), 1.27–1.29.
10. Kershner, *Student Pilot's Manual*, 100–103.
11. Adrian L. R. Thomas and Graham K. Taylor, "Animal Flight Dynamics. I. Stability in Gliding Flight," *Journal of Theoretical Biology* 212, no. 3 (2001): 399–424.
12. Abzug and Larrabee, *Airplane Stability and Control*, 65–66.
13. David E. Alexander, *Nature's Flyers: Birds, Insects, and the Biomechanics of Flight* (Baltimore: Johns Hopkins University Press, 2002), 124–27.
14. Adrian L. R. Thomas, "The Flight of Birds That Have Wings and a Tail: Variable Geometry Expands the Envelope of Flight Performance," *Journal of Theoretical Biology* 183 (1996): 237–46.

15. Adrian L.R. Thomas, "On the Aerodynamics of Birds' Tails," *Philosophical Transactions of the Royal Society of London B* 340 (1993): 361–80.
16. Thomas and Taylor, "Animal Flight Dynamics. I."
17. Charles H. Gibbs-Smith, *The Aeroplane: An Historical Survey of Its Origins and Development* (London: Her Majesty's Stationery Office, 1960), 10.
18. John D. Anderson, *The Airplane: A History of Its Technology* (Reston, Va: American Institute of Aeronautics and Astronautics, 2002), 356.
19. Karl Nickel and Michael Wohlfahrt, *Tailless Aircraft in Theory and Practice*, AIAA Education Series (Washington, D.C.: American Institute of Aeronautics and Astronautics, 1994), 276–77.
20. Ibid., 35–36, 279.
21. Ibid., 286, 303.
22. Fernando Galé, *Tailless Tale* (Olalla, Wash.: B2 Streamlines, 1991), 11, 18, 23.
23. Colin J. Pennycuick, *Animal Flight* (London: Arnold, 1972), 27.
24. Vance A. Tucker and G. Christian Parrott, "Aerodynamics of Gliding Flight in a Falcon and Other Birds," *Journal of Experimental Biology* 52 (1970): 345–67.
25. Other aviation pioneers knew of these remarkable seeds and their aerodynamic properties. Unfortunately for historians, as biologists learned more about the biology of Southeast Asia, they had to update and revise their ideas about the biology and relationships of many of the plants and animals. As a result, among many other taxonomic changes, *Zanonia* is now known as *Alsomitra*.
26. Akira Azuma and Yoshinori Okuno, "Flight of a Samara, *Alsomitra macrocarpa*," *Journal of Theoretical Biology* 129 (1987): 263–74.
27. E. T. Wooldridge, *Winged Wonders: The Story of the Flying Wings* (Washington, D.C.: Smithsonian Institution Press, 1983), 10–12.
28. Ibid., 31–33.
29. Ibid.
30. Ibid., 145–72.
31. Anderson, *The Airplane*, 357–58.

Chapter 6: Flight Instruments

1. Charles F. Spence, *Aeronautical Information Manual/Federal Aviation Regulations: AIM/FAR* (New York: McGraw-Hill, 2002), 77–78.
2. R. Fox, S. W. Lehmkuhle, and D. H. Westendorf, "Falcon Visual Acuity," *Science* 192, no. 4236 (1976): 263–65.
3. C. Marullo, J. G. Pilon, and M. Mouze, "Postembryonic Development of the Optic Lobe in *Lestes eurinus* Say and *Aeshna mixta* Latreille: Volumetric Growth (Zygoptera: Lestidae; Anisoptera: Aeshnidae)," *Odonatologica* 16, no. 4 (1987): 379–84.
4. D. C. Van Essen, C. H. Anderson, and D. J. Felleman, "Information Processing in the Primate Visual System: An Integrated Systems Perspective," *Science* 255, no. 5043 (1992): 419–23.
5. M. V. Srinivasan and S. W. Zhang, "Visual Motor Computations in Insects," *Annual Review of Neuroscience* 27 (2004): 679–96.
6. Miriam Lehrer, "Looking All Around: Honeybees Use Different Cues in Different Eye Regions," *Journal of Experimental Biology* 201, no. 24 (1998): 3275–92.
7. M. V. Srinivasan, S. W. Zhang, and N. J. Bidwell, "Visually Mediated Odometry in Honeybees," *Journal of Experimental Biology* 200, no. 19 (1997): 2513–22.
8. Srinivasan and Zhang, "Visual Motor Computations in Insects."
9. Knut Schmidt-Nielsen, *Animal Physiology: Adaptation and Environment*, 5th ed. (Cambridge, U.K.: Cambridge University Press, 1997), 542–44.
10. Ibid., 543.

11. Robert Buderi, *The Invention That Changed the World: How a Small Group of Radar Pioneers Won the Second World War and Launched a Technological Revolution*, Sloan Technology Series (New York: Simon and Schuster, 1996), 52–76.
12. David K. Barton and Sergey A. Leonov, *Radar Technology Encyclopedia* (Boston: Artech House, 1997), 320.
13. Ibid., 322–24, 327.
14. William K. Kershner, *The Student Pilot's Flight Manual: From First Flight to the Private Certificate*, 5th ed. (Ames: Iowa State University Press, 1979), 15–16.
15. Wolfgang Wiltschko and Roswitha Wiltschko, "Magnetic Compass of European Robins," *Science* 176 (1972): 62–64.
16. Wolfgang Wiltschko and Roswitha Wiltschko, "Magnetic Orientation in Birds," in *Current Ornithology*, ed. R. F. Johnston (New York: Plenum, 1988), 67–121.
17. James H. Doolittle and Carroll V. Glines, *I Could Never Be So Lucky Again* (New York: Bantam, 1991), 135–50.
18. Ed Sternstein and Todd Gold, *From Takeoff to Landing: Everything You Wanted to Know about Airplanes but Had No One to Ask* (New York: Pocket Books, 1991), 178–80.
19. Lawrence G. Goodman and William H. Warner, *Dynamics* (Belmont, Calif.: Wadsworth, 1963), 433–38.
20. Kershner, *Student Pilot's Manual*, 20.
21. Ibid., 17–18.
22. Ibid., 18–19.
23. Horst Mittelstaedt, "Physiologie des Glechgewichtssinnes Bei Fliegenden Libellen," *Zeitschrift für Vergleichende Physiologie* 32 (1950): 422–63.
24. J. W. S. Pringle, "The Gyroscopic Mechanism of the Halteres of Diptera," *Philosophical Transactions of the Royal Society of London B* 233 (1948): 347–84.
25. Michael H. Dickinson, "Haltere-Mediated Equilibrium Reflexes of the Fruit Fly, *Drosophila melanogaster*," *Philosophical Transactions of the Royal Society of London B* 354 (1999): 903–16.
26. Vincent G. Dethier, *To Know a Fly* (San Francisco: Holden-Day, 1962), 2.
27. James Wang, "Gyros for R/C Helicopters" (http://www.heliproz.com/jwgyros.html), accessed March 31, 2008.
28. Hidetaka Abe, "Applications Expand for Downsized Piezoelectric Vibrating Gyroscopes," *Asia Electronics Industry* (April 2003): 49–50.
29. Kershner, *Student Pilot's Manual*, 12–13.
30. Greg Miller, "Society for Neuroscience Meeting: Bats Have a Feel for Flight," *Science* 310, no. 5752 (2005): 1260–61.
31. Torkel Weis-Fogh, "Biology and Physics of Locust Flight. IV. Notes on Sensory Mechanisms in Locust Flight," *Philosophical Transactions of the Royal Society of London B* 239, no. 667 (1956): 553–84.
32. Michael Gewecke, "The Antennae of Insects As Air-Current Sense Organs and Their Relationship to the Control of Flight," in *Experimental Analysis of Insect Behavior*, ed. L. B. Browne (Berlin: Springer-Verlag, 1974), 100–113.
33. Miller, "Bats Feel Flight."
34. Spence, *AIM/FAR*, 78, 86.
35. Kershner, *Student Pilot's Manual*, 11–12.
36. Sternstein and Gold, *From Takeoff to Landing*, 64–65.
37. James D. Burke, *The Gossamer Condor and Albatross: A Case Study in Aircraft Design*, AIAA Professional Study Series (New York: American Institute of Aeronautics and Astronautics, 1980), 6.1.
38. Kershner, *Student Pilot's Manual*, 21.

Chapter 7: Dispensing with Power

1. Norman Ellison, *British Gliders and Sailplanes, 1922–1970* (New York: Barnes and Noble, 1971), 9.

2. Charles H. Gibbs-Smith, *Aviation: An Historical Survey from Its Origins to the End of World War II* (London: Her Majesty's Stationery Office, 1970), 209.

3. Don Dwiggins, *On Silent Wings: Adventures in Motorless Flight* (New York: Grossett and Dunlap, 1970), 69–85.

4. The glide ratio—the distance a glider moves forward for every foot of altitude lost in a glide—turns out to be the same as the lift-to-drag ratio. So the forward speed and the lift-to-drag ratio together set the sinking speed. David E. Alexander, *Nature's Flyers: Birds, Insects, and the Biomechanics of Flight* (Baltimore: Johns Hopkins University Press, 2002), 37–39.

5. D. S. Halacy, *With Wings As Eagles: The Story of Soaring* (Indianapolis: Bobbs-Merrill, 1975), 57.

6. Carle Conway, *The Joy of Soaring: A Training Manual* (Los Angeles: Soaring Society of America, 1969), 24–36.

7. Lewin B. Barringer, *Flight without Power: The Art of Gliding and Soaring* (New York: Pitman, 1940), 85–106.

8. Alexander, *Nature's Flyers*, 66.

9. Henk Tennekes, *The Simple Science of Flight: From Insects to Jumbo Jets* (Cambridge, Mass.: MIT Press, 1996), calculated from data in appendix.

10. Colin J. Pennycuick, "Mechanics of Flight," in *Avian Biology*, ed. D. S. Farner and J. R. King (New York: Academic Press, 1975), 1–75.

11. Halacy, *Wings As Eagles*, 37.

12. Conway, *Soaring*, 130.

13. Alexander, *Nature's Flyers*, 228–29.

14. Ibid., 59–60.

15. Conway, *Soaring*, 87–88; David L. Gibo, "Some Observations on Slope Soaring in *Pantala flavescens* (Odonata: Libellulidae)," *Journal of the New York Entomological Society* 89 (1981): 184–87.

16. Halacy, *Wings As Eagles*, 106.

17. Alexander, *Nature's Flyers*, 60–61.

18. Dwiggins, *Silent Wings*, 111–13.

19. Sankar Chatterjee, R. Jack Templin, and Kenneth E. Campbell, "The Aerodynamics of *Argentavis*, the World's Largest Flying Bird from the Miocene of Argentina," *Proceedings of the National Academy of Sciences* 104, no. 30 (2007): 12398–403.

20. Wann Langston, Jr., "Pterosaurs," *Scientific American* 224 (1981): 122–36.

21. Paul Alan Cox, "Observations on the Natural History of Samoan Bats," *Mammalia* 47 (1983): 519–23, and unpublished observations of E. D. Pierson and W. E. Rainey.

22. Alexander, *Nature's Flyers*, 62.

23. Vance A. Tucker, "Drag Reduction by Wing Tip Slots in a Gliding Harris' Hawk, *Parabuteo unicinctus*," *Journal of Experimental Biology* 198 (1995): 775–81.

24. Tennekes, *Simple Science*, appendix.

25. David L. Gibo and M. J. Pallett, "Soaring Flight of Monarch Butterflies, *Danaus plexippus* (Lepidoptera, Danaidae), during the Late Summer Migration in Southern Ontario," *Canadian Journal of Zoology* 57, no. 7 (1979): 1393–1401.

26. Colin J. Pennycuick, Thomas Alerstam, and Bertil Larsson, "Soaring Migration of the Common Crane *Grus grus* Observed by Radar and from an Aircraft," *Ornis Scandinavica* 10 (1979): 241–51.

27. Thomas Alerstam, "The Course and Timing of Bird Migration," in *Animal Migration*, ed. D. J. Aidley, Society for Experimental Biology Seminar Series (Cambridge, U.K.: Cambridge University Press, 1981), 9–54.
28. Alexander, *Nature's Flyers*, 59.
29. Halacy, *Wings As Eagles*, 160.
30. Barringer, *Without Power*, 5; Halacy, *Wings As Eagles*, 160.
31. Dwiggins, *Silent Wings*, 33–39.
32. Halacy, *Wings As Eagles*, 161.
33. Dwiggins, *Silent Wings*, 43–45; Halacy, *Wings As Eagles*, 161.
34. Halacy, *Wings As Eagles*, 163.
35. Barringer, *Without Power*, 13.
36. Halacy, *Wings As Eagles*, 166.
37. Fédération Aéronautique Internationale (FAI), "Gliding World Records" (http://records.fai.org/gliding/#current), accessed May 2, 2008.
38. Dwiggins, *Silent Wings*, 33–42.
39. Barringer, *Without Power*, 7–8.
40. Halacy, *Wings As Eagles*, 159.
41. Barringer, *Without Power*, 11.
42. Halacy, *Wings As Eagles*, 159.
43. Ibid., 160.
44. Barringer, *Without Power*, 9–11.
45. Halacy, *Wings As Eagles*, 175.
46. Dwiggins, *Silent Wings*, 111–15.
47. Phillip Aubrey Wills, *The Beauty of Gliding* (New York: Pitman, 1960), 19–23.
48. Halacy, *Wings As Eagles*, 176–77.
49. FAI, "Records."
50. Jane's Information Service, *Jane's All the World's Aircraft* (http://jawa.janes.com/docs/jawa/search.jsp), accessed March 31, 2008.
51. Wills, *The Beauty of Gliding*, 22.
52. FAI, "Records."
53. Ibid.
54. Barringer, *Without Power*, 36, 38.
55. Halacy, *Wings As Eagles*, 42; "Index of Sailplanes" (http://www.sailplanedirectory.com/ndxtype.htm), accessed May 15, 2008.
56. "Index of Sailplanes."
57. Richard Eppler, *Airfoil Design and Data* (New York: Springer-Verlag, 1990).
58. Ellison, *British Gliders*, 21.

Chapter 8: Straight Up

1. Knut Schmidt-Nielsen, *Scaling: Why Is Animal Size So Important?* (New York: Cambridge University Press, 1984), 164.
2. Steven Vogel, *Life in Moving Fluids: The Physical Biology of Flow*, 2d ed. (Princeton, N.J.: Princeton University Press, 1994), 264.
3. Charles Gablehouse, *Helicopters and Autogiros: A History of Rotating-Wing and V/STOL Aviation*, rev. ed. (Philadelphia: Lippincott, 1969), 4.
4. Jay P. Spenser, *Whirlybirds: A History of the U.S. Helicopter Pioneers* (Seattle: University of Washington Press, 1998), 187.
5. Frank K. Ross, *Flying Windmills: The Story of the Helicopter* (New York: Lee and Shephard, 1953), 111–13.
6. Gablehouse, *Helicopters and Autogiros*, 229–30.

7.   This gyroscopic effect was even overlooked by capable engineers working on Igor Sikorsky's first successful helicopter. It took them multiple tries to develop a cyclic control mechanism that actually tilted the rotor in the same direction that the pilot moved the stick. Spenser, *Whirlybirds*, 32.

8.   Ross, *Flying Windmills*, 114–16.

9.   J. Seddon, *Basic Helicopter Aerodynamics*, AIAA Education Series (Washington, D.C.: American Institute of Aeronautics and Astronautics, 1990), 2.

10.   Ross, *Flying Windmills*, 113–14.

11.   Charles P. Ellington, "The Aerodynamics of Hovering Insect Flight. I. The Quasisteady Analysis," *Philosophical Transactions of the Royal Society of London B* 305 (1984): 1–15.

12.   R. Åke Norberg, "Hovering Flight of the Dragonfly *Aeschna juncea* L., Kinematics and Aerodynamics," in *Swimming and Flying in Nature*, ed. T. Y. Wu, C. J. Brokaw, and C. Brennan (New York: Plenum, 1975), 763–81.

13.   Z. Jane Wang, "The Role of Drag in Insect Hovering," *Journal of Experimental Biology* 207, no. 23 (2004): 4147–55.

14.   C. C. Voigt and Y. Winter, "Energetic Cost of Hovering Flight in Nectar-Feeding Bats (Phyllostomidae: Glossophaginae) and Its Scaling in Moths, Birds and Bats," *Journal of Comparative Physiology B: Biochemical Systemic and Environmental Physiology* 169, no. 1 (1999): 38–48.

15.   Spenser, *Whirlybirds*, 440–44, 455–64.

16.   National Highway Traffic Safety Administration and American Medical Association Commission on Emergency Medical Services, *Air Ambulance Guidelines* (Washington, D.C: U.S. Department of Transportation and American Medical Association, 1986).

17.   Gablehouse, *Helicopters and Autogiros*, 117–18.

18.   "Aerospace Laureates: Operations," *Aviation Week and Space Technology* 146, no. 14 (1996): 16–17.

19.   Walter J. Boyne and Donald S. Lopez, *Vertical Flight: The Age of the Helicopter* (Washington, D.C.: Smithsonian Institution Press, 1984), 133–43.

20.   Jane's Information Service, *Jane's All the World's Aircraft* (http://jawa.janes.com/docs/jawa/search.jsp), accessed March 31, 2008.

21.   Spenser, *Whirlybirds*, 210, 300.

22.   Tom D. Crouch, *The Bishop's Boys: A Life of Wilbur and Orville Wright* (New York: Norton, 1989), 56–57.

23.   Charles H. Gibbs-Smith, *Aviation: An Historical Survey from Its Origins to the End of World War II* (London: Her Majesty's Stationery Office, 1970), 36.

24.   Ibid., 125.

25.   Gablehouse, *Helicopters and Autogiros*, 26–27.

26.   Spenser, *Whirlybirds*, 4–8.

27.   J. Gordon Leishman, *Principles of Helicopter Aerodynamics*, vol. 12, Cambridge Aerospace Series (New York: Cambridge University Press, 2000), 12–13.

28.   Gablehouse, *Helicopters and Autogiros*, 61–64.

29.   Ross, *Flying Windmills*, 69–74.

30.   Spenser, *Whirlybirds*, 12.

31.   Ibid., 20–34.

32.   Boyne and Lopez, *Vertical Flight*, 75–79.

33.   Spenser, *Whirlybirds*, 287, 308.

34.   Ibid., 304–8.

35.   Boyne and Lopez, *Vertical Flight*, 86–97.

36.   Spenser, *Whirlybirds*, 123–71.

37. Seddon, *Helicopter Aerodynamics*, 39.
38. David Rendall, *Jane's Aircraft Recognition Guide* (Glasgow: HarperCollins, 1996), 147, 185, 263.
39. Boyne and Lopez, *Vertical Flight*, 178–80.
40. Ibid., 183–89.
41. John Croft, "Tilters," *Air and Space Smithsonian* 22, no. 4 (2007): 40–47.
42. U.S. Congress, Senate Committee on Armed Services, *Report of the Panel to Review the V-22 Program: Hearing before the Committee on Armed Services, United States Senate, One Hundred Seventh Congress, First Session, May 1, 2001* (Washington, D.C.: U.S. Government Printing Office, 2002).
43. John W. Fozard, *The Jet V/STOL Harrier: An Evolutionary Revolution in Tactical Air Power: A Historical Account, from an Engineering Viewpoint, of the Development of the Harrier Family of Vectored Thrust V/STOL Attack Fighters,* AIAA Professional Study Series (Kingston upon Thames, Surrey, U.K.: British Aerospace, Aircraft Group, Kingston-Brough Division, 1978).

Chapter 9: Stoop of the Falcon

1. T. J. Cade, J. E. Enderson, C. G. Thelander, and C. M. White, eds., *Peregrine Falcon Populations: Their Management and Recovery* (Boise, Idaho: Peregrine Fund, 1988), 100, 403, 551–53, 561.
2. Ibid.
3. Derek A. Ratcliffe, *The Peregrine Falcon*, 2d ed. (London: Poyser, 1993), 12–19.
4. J. L. Osorno, R. Torres, and C. Macias Garcia, "Kleptoparasitic Behavior of the Magnificent Frigatebird: Sex Bias and Success," *Condor* 94, no. 3 (1992): 692–98.
5. A. G. Popa-Lisseanu, A. Delgado-Huertas, M. G. Forero, A. Rodríguez, R. Arlettaz, and C. Ibáñez, "Bats' Conquest of a Formidable Foraging Niche: The Myriads of Nocturnally Migrating Songbirds," *PLoS ONE* 2, no. 2 (2007): E205.
6. John Nielsen, "Giant Bats Snatch Birds from Night Sky," *Morning Edition* (Washington, D.C.: National Public Radio, April 18, 2007).
7. Gareth Jones and Jens Rydell, "Hawking and Gleaning," in *Bat Ecology*, ed. T. H. Kunz and M. B. Fenton (Chicago: University of Chicago Press, 2003), 301–45, 307–8.
8. Ibid., 307–10.
9. Klaus Richarz and Alfred Limbrunner, *The World of Bats: The Flying Goblins of the Night* (Neptune City, N.J.: T.F.H. Publications, 1993), 37–47.
10. Raleigh J. Robertson, Bridget Joan Stutchbury, and R. R. Cohen, *Tree Swallow: "Tachycineta bicolor,"* Birds of North America Series, no. 11 (Washington, D.C.: American Ornithologists' Union and Academy of Natural Sciences, 1992); C. R. Brown and Mary B. Brown, *Cliff Swallow: "Hirundo pyrrhonota,"* Birds of North America Series, no. 149 (Washington, D.C.: American Ornithologists' Union and Academy of Natural Sciences, 1995).
11. Paul D. Kyle and Georgean Z. Kyle, *Chimney Swifts: America's Mysterious Birds above the Fireplace,* Louise Lindsey Merrick Natural Environment Series, no. 37 (College Station: Texas A&M University Press, 2005).
12. R. G. Poulin, S. D. Grindal, and Robert Mark Brigham, *Common Nighthawk: "Chordeiles minor,"* Birds of North America Series, no. 213 (Washington, D.C.: American Ornithologists' Union and Academy of Natural Sciences, 1996).
13. Phil Chantler, *Swifts: A Guide to the Swifts and Treeswifts of the World*, 2d ed. (New Haven, Conn.: Yale University Press, 2000), 32–35, 41–42; David Lambert Lack, *Swifts in a Tower* (London: Methuen, 1956), 98–109; Brown and Brown, *Cliff Swallow*; Poulin et al., *Common Nighthawk*.

14. Howell V. Daly, John T. Doyen, and Alexander H. Purcell, III, *Introduction to Insect Biology and Diversity*, 2d ed. (Oxford: Oxford University Press, 1998), 234.
15. Richard J. Elzinga, *Fundamentals of Entomology*, 6th ed. (Upper Saddle River, N.J.: Pearson Education, 2004), 287.
16. Ibid., 263.
17. May Berenbaum, *Bugs in the System: Insects and Their Impact on Human Affairs* (Reading, Mass.: Addison-Wesley, 1995), 197.
18. Mark L. Winston, *The Biology of the Honey Bee* (Cambridge, Mass: Harvard University Press, 1987), 206–7.
19. S. M. Fitzpatrick and W. G. Wellington, "Contrasts in the Territorial Behavior of 3 Species of Hover Flies (Diptera: Syrphidae)," *Canadian Entomologist* 115, no. 5 (1983): 559–66.
20. Marc Wortman, *The Millionaires' Unit : The Aristocratic Flyboys Who Fought the Great War and Invented American Airpower*, 1st ed. (New York: PublicAffairs, 2006), 32, 40–41.
21. R. G. Grant, *Flight: 100 Years of Aviation* (New York: Dorling Kindersley, 2002), 68–72.
22. Ibid., 72–73.
23. The World War I Allies initially called fighter airplanes *scouts*, which is confusing because observation planes did most of what we think of as scouting. Similarly, the German term for fighter airplanes translates as "hunter." The U.S. Army officially called fighters *pursuit planes* or *pursuits* from the 1920s until World War II. In contrast, the U.S. Navy and the British Royal Air Force used the term *fighter* during that period. By the beginning of World War II, even though the U.S. Army Air Force was designating its fighters with "P" for *pursuit*—P-38 Lightning, P-51 Mustang—*fighter* had replaced *pursuit* even in official usage. Eric M. Bergerud, *Fire in the Sky: The Air War in the South Pacific* (Boulder, Colo.: Westview, 2000), 255.
24. John D. Anderson, *The Airplane: A History of Its Technology* (Reston, Va.: American Institute of Aeronautics and Astronautics, 2002), 131–37. Modern military terminology for categories of fighter airplanes is less than transparent. In World War II, fighters that could also carry a small load of bombs and attack ground targets were called fighter-bombers. Nowadays, such planes are called tactical fighters as opposed to interceptors (for attacking bombers) and air superiority fighters (for attacking fighters or bombers), which only attack other airplanes. Some recent tactical fighters—F-105 Thunderchief, F-111 Aardvark—are more bomber than fighter, which blurs the distinction between fighters and attack planes. Attack planes are maneuverable, one- or two-seat, combat airplanes intended solely for ground attack, typically the size of a fighter but not as fast. In this chapter, I am considering fighters only in their aerial combat role, not in their ground attack role.
25. James H. Doolittle and Carroll V. Glines, *I Could Never Be So Lucky Again* (New York: Bantam, 1991), 379–81.
26. Bergerud, *Fire in the Sky*, 231–32, 411.
27. Ibid., 200.
28. E. T. Wooldridge, *Into the Jet Age: Conflict and Change in Naval Aviation, 1945–1975: An Oral History* (Annapolis, Md.: Naval Institute Press, 1995), 227.
29. Mark A. Lorell and Hugh P. Levaux, *The Cutting Edge: A Half Century of Fighter Aircraft R&D* (Santa Monica, Calif.: RAND, 1998), 210; Robert K. Wilcox, *Scream of Eagles: The Creation of Top Gun and the U.S. Air Victory in Vietnam* (New York: Wiley, 1990), 96–106.
30. David Noland, "The Bone Is Back," *Air and Space Smithsonian* 23, no. 1 (2008): 60–67.
31. Bergerud, *Fire in the Sky*, 242.
32. Daly et al., *Insect Biology*, 135.

33. L. A. Miller and A. Surlykke, "How Some Insects Detect and Avoid Being Eaten by Bats: Tactics and Countertactics of Prey and Predator," *Bioscience* 51, no. 7 (2001): 570–81.
34. David K. Barton and Sergey A. Leonov, *Radar Technology Encyclopedia* (Boston: Artech House, 1997), 393.
35. Robert Buderi, *The Invention That Changed the World: How a Small Group of Radar Pioneers Won the Second World War and Launched a Technological Revolution*, Sloan Technology Series (New York: Simon and Schuster, 1996), 192–209.
36. Barton and Leonov, *Radar*, 74, 106.
37. Ibid., 225–31.
38. Ibid., 229–30.
39. Ibid., 228.
40. B. Mohl and L. A. Miller, "Ultrasonic Clicks Produced by the Peacock Butterfly: A Possible Bat Repellent Mechanism," *Journal of Experimental Biology* 64, no. 3 (1976): 639–44.
41. Ibid.
42. Nickolay I. Hristov and William E. Conner, "Sound Strategy: Acoustic Aposematism in the Bat–Tiger Moth Arms Race," *Naturwissenschaften* 92, no. 4 (2005): 164–69.
43. Bergerud, *Fire in the Sky*, 159.
44. Doolittle and Glines, *Lucky*, 103–7, 169–84.
45. Grant, *Flight*, 180.
46. Bergerud, *Fire in the Sky*, 262–63.
47. Anderson, *The Airplane*, 310–11.
48. Lorell and Levaux, *The Cutting Edge*, 91, 105.
49. Ibid., 91, 92, 104.
50. Not only does rolling inverted to start a dive speed up the dive, but it also improves the pilot's view downward and maintains positive g forces. If the pilot simply shoves the control stick forward, centrifugal force will be upward, and both pilot and aircraft will feel a reduction in weight or even an upward pull—negative g's—as they tip forward. Negative g's are hard on the pilot's body and the airplane's engine, so rolling inverted just at the start of the maneuver makes the centrifugal pull feel like the normal direction of gravity, or positive g's.
51. Vance A. Tucker, Tom J. Cade, and Alice E. Tucker, "Diving Speeds and Angles of a Gyrfalcon (*Falco rusticolus*)," *Journal of Experimental Biology* 201, no. 13 (1998): 2061–70.
52. Michael H. Dickinson, "Haltere-Mediated Equilibrium Reflexes of the Fruit Fly, *Drosophila melanogaster*," *Philosophical Transactions of the Royal Society of London B* 354 (1999): 903–16.

Chapter 10: Biology Meets Technology Head On

1. Peter Haining, *The Compleat Birdman: An Illustrated History of Man-Powered Flight* (New York: St. Martin's, 1976), 40–41.
2. Ibid., 62–63
3. Clive Hart, *The Prehistory of Flight* (Berkeley: University of California Press, 1985), 108–15.
4. John D. Anderson, *Inventing Flight: The Wright Brothers and Their Predecessors* (Baltimore: Johns Hopkins University Press, 2004), 9.
5. Charles H. Gibbs-Smith, *Aviation: An Historical Survey from Its Origins to the End of World War II* (London: Her Majesty's Stationery Office, 1970), 8.
6. Haining, *Compleat Birdman*, 102–3.

7. Gibbs-Smith, *Aviation*, 31–34; D. A. Reay, *The History of Man-Powered Flight* (New York: Pergamon, 1977), 35.
8. Gibbs-Smith, *Aviation*, 48.
9. Ibid., 31–34; Reay, *History*, 37, 47.
10. Otto Lilienthal, *Birdflight As the Basis of Aviation*, trans. A. W. Isenthal (Hummelstown, Penn.: Markowski International, 2001), 30–32.
11. Anderson, *Inventing Flight*, 68.
12. Reay, *History*, 58–59.
13. Ibid., 63.
14. Keith Sherwin, *Man-Powered Flight*, rev. ed. (Kings Langley, Hertfordshire, U.K.: Model and Allied Publications, 1975), 14.
15. Reay, *History*, 88–92.
16. D. R. Wilkie, "Muscle Function: A Personal View," *Journal of Experimental Biology* 115 (1985): 1–13.
17. D. R. Wilkie, "Man As a Source of Mechanical Power," *Ergonomics* 3, no. 1 (1960): 1–8; D. R. Wilkie, "Man As an Aero Engine," *Journal of the Royal Aeronautical Society* 64, no. 596 (1960): 477–81.
18. Thomas A. McMahon, *Muscles, Reflexes, and Locomotion* (Princeton, N.J.: Princeton University Press, 1984), 38–48.
19. Reay, *History*, 64–72.
20. Ibid., 92–98.
21. Ibid., 109–12.
22. Ibid., 138–41.
23. Ibid., 147.
24. Ibid., 154.
25. Morton Grosser, *Gossamer Odyssey: The Triumph of Human-Powered Flight* (Boston: Houghton Mifflin, 1981), 24–25.
26. Haining, *Compleat Birdman*, 125.
27. Chris Roper, "Human Powered Flying: True Flights" (http://www.humanpoweredfly-ing.propdesigner.co.uk/html/flights.html), accessed March 31, 2008.
28. Ibid.
29. Reay, *History*, 164–71, 182–184; Sherwin, *Man-Powered Flight*, 114–16.
30. Reay, *History*, 170–71, 184–186; Roper, "Human Powered Flying: True Flights"
31. Roper, "Human Powered Flying: True Flights"
32. Reay, *History*, 186–88.
33. Sherwin, *Man-Powered Flight*, 181; Chris Roper, "Human Powered Flying: Other '70's Planes" (http://www.humanpoweredflying.propdesigner.co.uk/html/other_70_s_planes.html), accessed March 31, 2008.
34. Reay, *History*.
35. Grosser, *Gossamer Odyssey*, 40.
36. Reay, *History*, 293–94; Chris Roper, "Human Powered Flying: Jupiter" (http://www. humanpoweredflying.propdesigner.co.uk/html/jupiter.html), accessed March 31, 2008.
37. Reay, *History*, 285–88.
38. Ibid., 288–90; Sherwin, *Man-Powered Flight*, 180–82.
39. Haining, *Compleat Birdman*, 124.
40. Reay, *History*, 296–98.
41. Grosser, *Gossamer Odyssey*, 66–70.
42. Ibid., 70–71, 74.

43. James D. Burke, *The Gossamer Condor and Albatross: A Case Study in Aircraft Design*, AIAA Professional Study Series (New York: American Institute of Aeronautics and Astronautics, 1980), 2.1.

44. Grosser, *Gossamer Odyssey*, 79–83; Burke, *Condor and Albatross*, 4.4.

45. Burke, *Condor and Albatross*, 2.1, 4.4.

46. Ibid., 4.4

47. Grosser, *Gossamer Odyssey*, 87–88.

48. Burke, *Condor and Albatross*, 2.3–2.4, 4.7, 5.4.

49. Ibid., 3.1, 4.5.

50. Ibid., 4.9; Grosser, *Gossamer Odyssey*, 114–16.

51. Burke, *Condor and Albatross*, 4.6.

52. Grosser, *Gossamer Odyssey*, 109.

53. Burke, *Condor and Albatross*, 4.9–4.11.

54. Grosser, *Gossamer Odyssey*, 122, 130.

55. Ibid., 138–45.

56. Chris Roper, "Human Powered Flying: The Gossamers and Other Planes" (http://www.humanpoweredflying.propdesigner.co.uk/html/the_gossamers_and_other_planes.html), accessed March 31, 2008; Burke, *Condor and Albatross*, 5.7–5.9.

57. Grosser, *Gossamer Odyssey*, 175–79, 182–84.

58. Ibid., 200.

59. Ibid., 201–7.

60. Ibid., 209–26.

61. Ibid., 240–53; Burke, *Condor and Albatross*, 7.1.

62. Chris Roper, "Human Powered Flying: Kremer Speed" (http://www.humanpoweredflying.propdesigner.co.uk/html/kremer_speed.html), accessed March 31, 2008.

63. Ibid.

64. Ibid.

65. In the Greek myth, Daedalus made wings out of feathers attached by wax for himself and his son Icarus. They used the wings to escape from captivity on Crete by flying across the Aegean Sea. Daedalus warned Icarus not to fly too close to the sun or the wax would melt, but Icarus, in his youthful exuberance, ignored the warning. The wax melted, Icarus fell to his death, and Daedalus flew on and crossed the sea alone.

66. Chris Roper, "Human Powered Flying: Daedalus" (http://www.humanpoweredflying.propdesigner.co.uk/html/daedalus.html), accessed March 31, 2008.

67. John S. Langford, "Triumph of *Daedalus*," *National Geographic* 174, no. 2 (1988): 191–99.

68. Roper, "Human Powered Flying: Daedalus"

69. Langford, "Triumph of *Daedalus*."

70. Ibid.; Roper, "Human Powered Flying: Daedalus."

71. Bruno Lange, *Typenhandbuch der Deutschen Luftfahrttechnik* (Koblenz, Germany: Bernard and Graefe Verlag, 1986), 229.

72. Nathan Chronister, "The Ornithopter Zone: Manned Ornithopter Flights" (http://www.ornithopter.org/manned.shtml), accessed March 31, 2008.

73. Nathan Chronister, "The Ornithopter Zone" (http://www.ornithopter.org/index.shtml), accessed March 31, 2008.

74. Henry R. Jex, "Making Pterodactyls Fly (QN Story)" (http://www.twitt.org/QNStory.html), accessed May 12, 2008; Paul B. MacCready, "The Great Pterodactyl Project," *Engineering and Science* 49 (1985): 18–24.

75. Sankar Chatterjee and R. Jack Templin, "Posture, Locomotion, and Paleoecology of Pterosaurs," *Geological Society of America Special Papers* 379 (2004): 1–64.

76. Paul B. MacCready, "Art and Nature in the Technology of Flight," *Leonardo* 19, no. 4 (1986): 317–19.
77. James D. DeLaurier and J. M. Harris, "A Study of Mechanical Flapping-Wing Flight," *Aeronautical Journal* 97 (1993): 277–86.
78. James D. DeLaurier, "The Development of an Efficient Ornithopter Wing," *Aeronautical Journal* 97, no. 965 (1993): 153–62.
79. Graham Chandler, "Ready, Set, Flap!" *Air and Space Smithsonian* 16, no. 5 (2002): 36–41.
80. Treena Hein, "The World's First Flying Ornithopter," *CBC News: In Depth* (Toronto: Canadian Broadcasting Corporation, October 12, 2006).
81. Matthew T. Keennon and Joel M. Grasmeyer, "Development of the Black Widow and Microbat MAVs and a Vision of the Future of MAV Design," in *AIAA/ICAS International Air and Space Symposium and Exposition: The Next 100 Years* (Dayton, Ohio: American Institute of Aeronautics and Astronautics, 2003).
82. Peter Garrison, "Microspies," *Air and Space Smithsonian* 15, no. 1 (2000): 54–61.
83. Robert Michelson, "Entomopter Project" (http://avdil.gtri.gatech.edu/RCM/RCM/Entomopter/EntomopterProject.html), accessed March 31, 2008.
84. Ron S. Fearing, "Micromechanical Flying Insect (MFI) Project" (http://robotics.eecs.berkeley.edu/~ronf/MFI/index.html), accessed March 30, 2008; R. J. Wood, S Avadhanula, M. Menon, and R. S. Fearing, "Microrobotics Using Composite Materials: The Micromechanical Flying Insect Thorax," in *Proceedings of the IEEE International Conference on Robotics and Automation* (Piscataway, N.J.: IEEE Press, 2003), 1842–49.

Epilogue

1. Robert Dudley, "Atmospheric Oxygen, Giant Paleozoic Insects and the Evolution of Aerial Locomotor Performance," *Journal of Experimental Biology* 201 (1998): 1043–50.
2. Stephen J. Gould and Niles Eldredge, "Punctuated Equilibria: The Tempo and Mode of Evolution Reconsidered," *Paleobiology* 3, no. 2 (1977): 115–51.

GLOSSARY

adverse yaw—a tendency for the nose of an aircraft in a bank to swing in the direction opposite to the bank due to unequal drag on the right and left ailerons.

aileron—a hinged, movable, flaplike control surface on the trailing edge of a wing near each tip. The left and right ailerons typically move in opposite directions and are used to control roll, or angle of bank. Ailerons are the primary control used for turning.

Airbus—originally a consortium of several European companies formed to build airliners. Now it is a multinational subsidiary of the European Aeronautic Defense and Space conglomerate and one of only two companies in the world currently building large airliners.

airfoil—the shape of a cross-section through the wing from front to back, usually rounded or blunt in front and sharply tapering at the back.

airliner—an airplane primarily designed to carry passengers.

angle of attack—the angle that the chord of a wing makes with the oncoming air. It measures how much the wing is tilted leading-edge-up relative to its direction of travel.

antagonistic muscles—two muscles or sets of muscles that move a joint in opposite directions, as in upstroke and downstroke muscles.

Archaeopteryx—shortened name for *Archaeopteryx lithographica*, an extinct species of feathered, flying animal from the Jurassic with both birdlike and reptilian traits. A number of existing fossils show fine details of its feathers and other soft body parts.

aspect ratio—for a rectangular wing, the ratio of the span to the chord (for other wing shapes, the ratio of the span squared to the planform area, which is mathematically equivalent). The aspect ratio is a measure of how narrow or broad a wing is.

bank—when a flyer is tilted or rolled to one side. By banking, a flyer tilts its lift to one side, which produces a turn.

biplane—an airplane with two sets of wings, one above the other.

Boeing—a large aerospace company, builder of many famous military airplanes and airliners. After several mergers, it is now the only U.S. company still building airliners. Boeing built the first commercially successful jet airliner (the 707), the most popular jet airliner (the 737), and the first and, until very recently, the largest jumbo jet (the 747).

camber—the upward convexity of an airfoil, usually measured as the maximum height of the airfoil centerline above the chord divided by the chord length.

canard—an airplane with horizontal control surfaces at the front rather than the back; a "tail-first" design. The term (from the French word for "duck") may also refer to the horizontal control surfaces of such a craft.

cantilever wing—a wing without the external bracing of wires or struts.

chord—a line connecting the forward-most point of a wing or an airfoil with the far-thest rearward point in a plane parallel to the wing's direction of movement or an animal's or an airplane's long axis. The term may also refer to the length of such a line.

collective pitch—the helicopter control that increases or decreases the average pitch of the rotor. It is used to adjust the amount of lift.

composite material—any material made by combining two substances with very different mechanical properties. In modern engineering, it usually refers to a fibrous material embedded in a stiff plastic, such as fiberglass or graphite-epoxy.

cyclic pitch—the helicopter control that causes the plane of the rotor to tilt by varying the pitch as the rotor rotates. It performs functions similar to those of an airplane's ailerons and elevator.

dihedral—having the wing tips elevated above the root so that they form a shallow V. The term may also refer to the angle of such a wing above the horizontal.

drag—a force parallel with the direction of movement or air flow but pushing in the opposite direction; a retarding force.

echolocation—the process of producing sounds and listening for echoes to determine distance and direction to solid objects. Used by bats and porpoises, it is also called sonar.

elevator—a hinged, movable, flaplike control surface on the trailing edge of the horizontal tail of an airplane that is used to control pitch and angle of attack.

factor of safety—the excess strength designed into a structure beyond the expected maximum load. For example, a factor of safety of two means the structure is designed to withstand twice the expected maximum load.

fighter—a military airplane designed primarily to attack other airplanes.

flapping—up-and-down wing movements used by flying animals to generate thrust.

flaps—movable control surfaces on the trailing edge (or sometimes the leading edge) of a wing that increase the camber when deflected. They allow an airplane to fly more slowly or descend more steeply, mainly for landing.

fly-by-wire—an aircraft control system in which the pilot's controls are connected to a set of computers that move the control surfaces using electrical signals in wires running to actuators at the control surfaces. The system usually includes automatic computer stabilization.

glide—unpowered flight in which the glider must descend (relative to the air) to maintain a steady speed.

glider—any vehicle or animal that flies without power.

glide ratio—the horizontal distance that a glider travels for every unit distance of vertical descent. The glide ratio is equivalent in value to the lift-to-drag ratio.

hover—flight with little or no forward motion.

halteres—gyroscopic sense organs found only in true flies (Diptera). Evolved from modified hindwings, halteres are tiny structures used to detect pitch, roll, and, to a lesser extent, yaw.

induced drag—drag produced by the same process that produces lift. A wing only produces induced drag when it is producing lift.

jet engine—an engine that produces thrust by burning fuel continuously, causing gas to expand and rush out through the exhaust. Practical jet engines such as turbojets and turbofans are all forms of gas turbines.

jumbo jet—any airliner that is wide enough to have two aisles running down the length of its cabin. Originally applied only to the Boeing 747 (the first jumbo jet), the term now refers to all such airliners and is synonymous with *widebody*. Jumbo jets include mini-jumbos such as the Boeing 767 and superjumbos such as the Airbus A380.

landing gear—wheels or skids for supporting an aircraft on the ground. They may be fixed or retractable and are known in the United Kingdom as the *undercarriage*.

leading edge—the front edge or margin of a wing or the front-most point of an airfoil.

lift—a force perpendicular to a wing's motion through the air or, equivalently, to the direction of oncoming air flowing over the wing. Lift is not necessarily perfectly vertical but usually involves an upward component to offset a flyer's weight.

lift-to-drag ratio—the ratio of the forces of lift and drag on a wing. The ratio is dimensionless and is an important figure of merit for the effectiveness of a wing or a flyer.

maneuverability—the ability to change directions rapidly.

micro air vehicle—an autonomous or radio-controlled flying machine with a wingspan of six inches (fifteen centimeters) or less.

monoplane—airplane with a single set of wings, the conventional wing arrangement of modern airplanes.

ornithopter—a flying machine that produces thrust by flapping its wings like a bird.

piston engine—an engine that works by detonating fuel in a cylinder to push down a piston, which turns a crankshaft. Such engines may also be called *reciprocating engines*.

pitch—nose-up or -down rotation about a horizontal axis parallel to the wings.

planform—the area defined by a wing's outline as viewed from directly above.

power—the rate of doing work as measured in horsepower or watts. Power is calculated either as work divided by time or as force times speed.

primary feathers—the largest flight feathers, or pinions, which form the tip of a bird's wing.

propeller—a set of blades on a shaft that function as rotating wings. When the shaft is rotated (as by a piston or a turboprop engine), the blades produce lift parallel to the shaft, which functions as thrust.

pterosaur—a member of a group of extinct flying reptiles related to dinosaurs and crocodiles, including the largest flying animals that ever lived as well as species as small as sparrows. Originally they were called *pterodactyls*, but that term now refers to a small subset of species of pterosaurs.

Reynolds number—the ratio of inertial (or pressure) forces and viscous forces, often used as a scale factor in aerodynamics. Small, slow flyers have low Reynolds numbers; large, fast flyers have high ones. The air seems more viscous to flyers with low Reynolds numbers such as insects, and they have lower lift-to-drag ratios. The term was named after pioneering hydrodynamicist Osborne Reynolds.

rib—a supporting structural element in a wing that runs from the leading edge to the trailing edge and defines the airfoil shape.

roll—rotation about a longitudinal axis; tilting to the right or the left.

rotor—a large set of rotating blades that produce lift to support a helicopter's weight. The term may also refer to the dangerous turbulence on the downwind side of a ridge used for slope soaring.

rudder—a hinged, movable, flaplike control surface on the trailing edge of a vertical stabilizer. It is used to control the left-right rotation, or yaw, of an airplane, although it is not usually the primary control for turning.

safety factor—see *factor of safety*.

sailplane—an engineless airplane (or glider) designed to stay aloft by taking advantage of rising air. Usually it has very long, high-aspect-ratio wings and a high lift-to-drag ratio.

samara—a seed with a wing, such as a maple seed, that is capable of gliding. Botanically, many samaras are actually fruits rather than seeds.

separated primaries—an arrangement of the primary feathers of a bird's wing that involves a slight separation between successive primary feathers. Many soaring birds used separated primaries to increase their lift-to-drag ratios.

sideslip—a situation in which an airplane's longitudinal axis is to the right or left of its direction of travel. Pilots avoid sideslips when turning but sometimes use them intentionally when landing in crosswinds or trying to make steeper descents.

slope soaring—soaring by taking advantage of wind blowing up a slope; sometimes called *declivity soaring*.

slotted wingtips—see *separated primaries*.

soaring—using rising air or other atmospheric energy to remain aloft while gliding.

sonar—acronym for "sound navigation and ranging." See *echolocation*.

span—the length of a wing from tip to tip, measured perpendicular to the body's long axis. Although biologists sometimes use the same term to describe the distance from a wingtip to the wing base, that distance is more appropriately called the *semispan*.

spoilers—control surfaces on top of a wing that can rise and "spoil" its lift. Sailplanes use spoilers to steepen their descent during landing, and most large airliners use them to prevent bouncing or inadvertent ascent after touchdown. If raised on one wing at a time, spoilers can be used as ailerons.

stability—a tendency to return to one's original heading after a disturbance causes a deviation from that heading.

stall—a situation that occurs when a wing is tilted to an angle of attack above some critical angle and the airflow separates from the upper surface of the wing, leaving a large, turbulent wake. When this happens to a wing at a high Reynolds number (such as those of birds and other relatively large flyers), lift drops and drag rises abruptly and dramatically. For wings at lower Reynolds numbers (such as those of insects), the lift decrease and drag increase are more gradual.

thermal—air that rises because it is warmer than surrounding air.

thrust—a force parallel to the direction of movement and in the same direction; a force that tends to maintain forward movement.

trailing edge—the rear edge or margin of a wing or the rearmost point of an airfoil.

turbofan—a jet engine with a compressor and a turbine in which some of the air passing through the compressor is routed directly to the exhaust, bypassing the combustion chamber and turbine. All modern jet engines are turbofans.

turbojet—a jet engine with a compressor and a turbine in which all air through the compressor goes to the combustion chamber and turbine. This was the earliest form of jet engine but is now obsolete.

turboprop—a jet engine with the turbine connected to a gearbox that turns a propeller.

variometer—a sensitive rate-of-climb indicator used in sailplanes.

VTOL—an abbreviation for "vertical take-off and landing." It refers to a craft that can take off and land vertically without a runway.

wingloading—total weight divided by planform area; weight per unit of planform area.

wing warping—the system invented by the Wright brothers for control of roll. It involves twisting the wings so the trailing edge at one wing tip is lowered and the trailing edge at the other tip is raised. Wing warping performs the same function as ailerons do.

yaw—nose-left or -right rotation about a vertical axis.

BIBLIOGRAPHY

Abbott, I. H. A., and A. E. von Doenhoff. 1959 [1949]. *Theory of Wing Sections, Including a Summary of Airfoil Data.* Reprint. New York: Dover.

Abe, H. 2003. "Applications Expand for Downsized Piezoelectric Vibrating Gyroscopes." *Asia Electronics Industry* (April): 49–50.

Abzug, M. J., and E. E. Larrabee. 1997. *Airplane Stability and Control: A History of the Technologies That Made Aviation Possible.* Cambridge Aerospace Series. Cambridge, U.K.: Cambridge University Press.

———. 2002. *Airplane Stability and Control: A History of the Technologies That Made Aviation Possible.* 2d ed. Cambridge Aerospace Series. Cambridge, U.K.: Cambridge University Press.

"Aerospace Laureates: Operations." 1996. *Aviation Week and Space Technology* 146, no. 14: 16–17.

Aldridge, H. D. J. N. 1987. "Body Accelerations during the Wingbeat in Six Bat Species: The Function of the Upstroke in Thrust Generation." *Journal of Experimental Biology* 130: 275–94.

Alerstam, T. 1981. "The Course and Timing of Bird Migration." In *Animal Migration*, edited by D. J. Aidley, pp. 9–54. Cambridge, U.K.: Cambridge University Press.

Alexander, D. E. 1986. "Wind Tunnel Studies of Turns by Flying Dragonflies." *Journal of Experimental Biology* 122: 81–98.

———. 2002. *Nature's Flyers: Birds, Insects, and the Biomechanics of Flight.* Baltimore: Johns Hopkins University Press.

Anderson, J. D. 2002. *The Airplane: A History of Its Technology.* Reston, Va: American Institute of Aeronautics and Astronautics.

———. 2004. *Inventing Flight: The Wright Brothers and Their Predecessors.* Baltimore: Johns Hopkins University Press.

Azuma, A., and Y. Okuno. 1987. "Flight of a Samara, *Alsomitra macrocarpa.*" *Journal of Theoretical Biology* 129: 263–74.

Barringer, L. B. 1940. *Flight without Power: The Art of Gliding and Soaring.* New York: Pitman.

Barton, D. K., and S. A. Leonov. 1997. *Radar Technology Encyclopedia.* Boston: Artech House.

Berenbaum, M. 1995. *Bugs in the System: Insects and Their Impact on Human Affairs.* Reading, Mass.: Addison-Wesley.

Bergerud, E. M. 2000. *Fire in the Sky: The Air War in the South Pacific.* Boulder, Colo.: Westview.

Bertin, J. J., and M. L. Smith. 1979. *Aerodynamics for Engineers.* Englewood Cliffs, N.J.: Prentice Hall.

Biewener, A. A., and K. P. Dial. 1995. "In Vivo Strain in the Humerus of Pigeons (*Columba livia*) during Flight." *Journal of Morphology* 225: 61–75.

Borrell, B. J., and M. J. Medeiros. 2004. "Thermal Stability and Muscle Efficiency in Hovering Orchid Bees (Apidae: Euglossini)." *Journal of Experimental Biology* 207, no. 17: 2925–33.

Boyne, W. J., and D. S. Lopez. 1984. *Vertical Flight: The Age of the Helicopter*. Washington, D.C.: Smithsonian Institution Press.

Bridgman, L. 1946. *Jane's All the World's Aircraft, 1945–46*. Vol. 34. New York: Macmillan.

Brodsky, A. K. 1994. *The Evolution of Insect Flight*. Oxford: Oxford University Press.

Brown, C. R., and M. B. Brown. 1995. *Cliff Swallow: "Hirundo pyrrhonota."* Birds of North America Series, no. 149. Washington, D.C.: American Ornithologists' Union and Academy of Natural Sciences.

Bryan, C. D. B. 1988. *The National Air and Space Museum*. 2d ed. New York: Abrams.

Buderi, R. 1996. *The Invention That Changed the World: How a Small Group of Radar Pioneers Won the Second World War and Launched a Technological Revolution*. Sloan Technology Series. New York: Simon and Schuster.

Bundle, M. W., K. S. Hansen, and K. P. Dial. 2007. "Does the Metabolic Rate-Flight Speed Relationship Vary among Geometrically Similar Birds of Different Mass?" *Journal of Experimental Biology* 210, no. 6: 1075–83.

Burke, J. D. 1980. *The Gossamer Condor and Albatross: A Case Study in Aircraft Design*. AIAA Professional Study Series. New York: American Institute of Aeronautics and Astronautics.

Cade, T. J., J. E. Enderson, C. G. Thelander, and C. M. White, eds. 1988. *Peregrine Falcon Populations: Their Management and Recovery*. Boise, Idaho: Peregrine Fund.

Chai, P., J. S. C. Chen, and R. Dudley. 1997. "Transient Hovering Performance of Hummingbirds under Conditions of Maximal Loading." *Journal of Experimental Biology* 200: 921–29.

Chandler, G. 2002. "Ready, Set, Flap!" *Air and Space Smithsonian* 16, no. 5: 36–41.

Chantler, P. 2000. *Swifts: A Guide to the Swifts and Treeswifts of the World*. 2d ed. New Haven, Conn.: Yale University Press.

Chatterjee, S., and R. J. Templin. 2004. "Posture, Locomotion, and Paleoecology of Pterosaurs." *Geological Society of America Special Papers* 379: 1–64.

Chatterjee, S., R. J. Templin, and K. E. Campbell. 2007. "The Aerodynamics of *Argentavis*, the World's Largest Flying Bird from the Miocene of Argentina." *Proceedings of the National Academy of Sciences* 104, no. 30: 12398–403.

Chiappe, L. M., J. Shu'An, J. Qiang, and M. A. Norell. 1999. "Anatomy and Systematics of the Confuciusornithidae (Theropoda: Aves) from the Late Mesozoic of Northeastern China." *Bulletin of the American Museum of Natural History* 242: 1–86.

Chronister, N. n.d. "The Ornithopter Zone." http://www.ornithopter.org/index.shtml. Accessed March 31, 2008.

———. n.d. "The Ornithopter Zone: Manned Ornithopter Flights." http://www.ornithopter.org/manned.shtml. Accessed March 31, 2008.

Conway, C. 1969. *The Joy of Soaring: A Training Manual*. Los Angeles: Soaring Society of America.

Cox, P. A. 1983. "Observations on the Natural History of Samoan Bats." *Mammalia* 47: 519–23.

Croft, J. 2007. "Tilters." *Air and Space Smithsonian* 22, no. 4: 40–47.

Crouch, T. D. 1989. *The Bishop's Boys: A Life of Wilbur and Orville Wright*. New York: Norton.

Crouch, T. D. 1998. "The Thrill of Invention." *Air and Space Smithsonian* 13, no. 1: 22–30.

Culick, F. E. C., and H. R. Jex. 1987. "Aerodynamics, Stability and Control of the 1903 Wright Flyer." In *The Wright Flyer: An Engineering Perspective*, edited by H. S. Wolko. Washington, D.C.: Smithonian Institution Press.

Currey, J. D. 1984. *The Mechanical Adaptations of Bones*. Princeton, N.J.: Princeton University Press.

Daly, H. V., J. T. Doyen, and A. H. Purcell, III. 1998. *Introduction to Insect Biology and Diversity*. 2d ed. Oxford: Oxford University Press.

"De Havilland Canada D.H.C.2 'Beaver I.'" n.d. *Aviation Enthusiast's Corner*. http://www.aero-web.org/specs/dehacana/dhc2i.htm. Accessed April 16, 2008.

DeLaurier, J. D. 1993. "The Development of an Efficient Ornithopter Wing." *Aeronautical Journal* 97, no. 965: 153–62.

DeLaurier, J. D., and J. M. Harris. 1993. "A Study of Mechanical Flapping-Wing Flight." *Aeronautical Journal* 97, no. 968: 277–86.

Dethier, V. G. 1962. *To Know a Fly*. San Francisco: Holden-Day.

Dial, K. P., A. A. Biewener, B. W. Tobalske, and D. R. Warrick. 1997. "Mechanical Power Output of Bird Flight." *Nature* 390, no. 6,655: 67–70.

Dickinson, M. H. 1999. "Haltere-Mediated Equilibrium Reflexes of the Fruit Fly, *Drosophila melanogaster*." *Philosophical Transactions of the Royal Society of London B* 354: 903–16.

Doolittle, J. H., and C. V. Glines. 1991. *I Could Never Be So Lucky Again*. New York: Bantam.

Downie, D., and J. Downie. 1987. *The Complete Guide to Rutan Aircraft*. 3d ed. Blue Ridge Summit, Penn.: Tab Books.

Dudley, R. 1998. "Atmospheric Oxygen, Giant Paleozoic Insects and the Evolution of Aerial Locomotor Performance." *Journal of Experimental Biology* 201: 1043–50.

Dudley, R., and C. P. Ellington. 1990. "Mechanics of Forward Flight in Bumblebees. II. Quasisteady Lift and Power Requirements." *Journal of Experimental Biology* 148: 53–88.

Dwiggins, D. 1970. *On Silent Wings: Adventures in Motorless Flight*. New York: Grossett and Dunlap.

Eggert, A. K., and S. K. Sakaluk. 1994. "Sexual Cannibalism and Its Relation to Male Mating Success in Sagebrush Crickets, *Cyphoderris strepitans* (Haglidae, Orthoptera)." *Animal Behaviour* 47, no. 5: 1171–77.

Ellington, C. P. 1984. "The Aerodynamics of Hovering Insect Flight. I. The Quasisteady Analysis." *Philosophical Transactions of the Royal Society of London* 305: 1–15.

Ellison, N. 1971. *British Gliders and Sailplanes, 1922–1970*. New York: Barnes and Noble.

Elzinga, R. J. 2004. *Fundamentals of Entomology*. 6th ed. Upper Saddle River, N.J.: Pearson Education.

Eppler, R. 1990. *Airfoil Design and Data*. New York: Springer-Verlag.

Fearing, R. S. n.d. "Micromechanical Flying Insect (MFI) Project." http://robotics.eecs.berkeley.edu/~ronf/MFI/index.html. Accessed March 30, 2008.

Fédération Aéronautique Internationale. n.d. "Gliding World Records." http://records.fai.org/gliding/#current. Accessed May 2, 2008.

Feduccia, A. 1999. *The Origin and Evolution of Birds*. 2d ed. New Haven, Conn.: Yale University Press.

Fitzpatrick, S. M., and W. G. Wellington. 1983. "Contrasts in the Territorial Behavior of 3 Species of Hover Flies Diptera Syrphidae." *Canadian Entomologist* 115, no. 5: 559–66.

Fox, R., S. W. Lehmkuhle, and D. H. Westendorf. 1976. "Falcon Visual Acuity." *Science* 192, no. 4,236: 263–65.

Fozard, J. W. 1978. *The Jet V/STOL Harrier: An Evolutionary Revolution in Tactical Air Power: A Historical Account, from an Engineering Viewpoint, of the Development of the Harrier Family of Vectored Thrust V/STOL Attack Fighters*. AIAA Professional Study Series. Kingston upon Thames, Surrey, U.K.: British Aerospace, Aircraft Group, Kingston-Brough Division.

Gablehouse, C. 1969. *Helicopters and Autogiros: A History of Rotating-Wing and V/STOL Aviation*. Revised edition. Philadelphia: Lippincott.

Galé, F. 1991. *Tailless Tale*. Olalla, Wash.: B2 Streamlines.

Garner, J. P., G. K. Taylor, and A. L. R. Thomas. 1999. "On the Origins of Birds: The Sequence of Character Acquisition in the Evolution of Avian Flight." *Proceedings of the Royal Society of London 266*, series B: 1259–66.

Garrison, P. 2000. "Microspies." *Air and Space Smithsonian* 15, no. 1: 54–61.

Gewecke, M. 1974. "The Antennae of Insects As Air-Current Sense Organs and Their Relationship to the Control of Flight." In *Experimental Analysis of Insect Behavior*, edited by L. B. Browne, pp. 100–113. Berlin: Springer-Verlag.

Gibbs-Smith, C. H. 1960. *The Aeroplane: An Historical Survey of Its Origins and Development*. London: Her Majesty's Stationery Office.

———. 1970. *Aviation: An Historical Survey from Its Origins to the End of World War II*. London: Her Majesty's Stationery Office.

Gibo, D. L. 1981. "Some Observations on Slope Soaring in *Pantala flavescens* (Odonata: Libellulidae)." *Journal of the New York Entomological Society* 89: 184–87.

Gibo, D. L., and M. J. Pallett. 1979. "Soaring Flight of Monarch Butterflies, *Danaus plexippus* (Lepidoptera, Danaidae), during the Late Summer Migration in Southern Ontario." *Canadian Journal of Zoology* 57, no. 7: 1393–401.

Gillmer, T. C., and B. Johnson. 1982. *Introduction to Naval Architecture*. Annapolis, Md.: Naval Institute Press.

Goebel, G. n.d. "Unmanned Aerial Vehicles." http://www.vectorsite.net/twuav_01.html#m1. Accessed April 30, 2008.

Goodman, L. G., and W. H. Warner. 1963. *Dynamics*. Belmont, Calif.: Wadsworth.

Gould, S. J., and N. Eldredge. 1977. "Punctuated Equilibria: The Tempo and Mode of Evolution Reconsidered." *Paleobiology* 3, no. 2: 115–51.

Grant, R. G. 2002. *Flight: 100 Years of Aviation*. New York: Dorling Kindersley.

Grosser, M. 1981. *Gossamer Odyssey: The Triumph of Human-Powered Flight*. Boston: Houghton Mifflin.

Gunston, B. 1998. *World Encyclopaedia of Aero Engines : All Major Aircraft Power Plants, from the Wright Brothers to the Present Day*. 4th ed. Newbury Park, Calif.: Haynes North America.

Haining, P. 1976. *The Compleat Birdman: An Illustrated History of Man-Powered Flight*. New York: St. Martin's.

Halacy, D. S. 1975. *With Wings As Eagles: The Story of Soaring*. Indianapolis: Bobbs-Merrill.

Hart, C. 1985. *The Prehistory of Flight*. Berkeley: University of California Press.

Hein, T. 2006. "The World's First Flying Ornithopter." *CBC News: In Depth*. Toronto: Canadian Broadcasting Corporation, October 12.

Herzog, K. 1968. *Anatomie und Flugbiologie der Vögel*. Stuttgart, Germany: Fischer.

Hoerner, S. F., and H. V. Borst. 1975. *Fluid-Dynamic Lift*. Bricktown, N.J.: Hoerner Fluid Dynamics.

Hoikkala, A., and P. Welbergen. 1995. "Signals and Responses of Females and Males in Successful and Unsuccessful Courtships of Three Hawaiian Lek-Mating *Drosophila* Species." *Animal Behaviour* 50, no. 1: 177–90.

Hristov, N. I., and W. E. Conner. 2005. "Sound Strategy: Acoustic Aposematism in the Bat–Tiger Moth Arms Race." *Naturwissenschaften* 92, no. 4: 164–69.

"Index of Sailplanes." n.d. http://www.sailplanedirectory.com/ndxtype.htm. Accessed May 15, 2008.

Jakab, P. L. 1990. *Visions of a Flying Machine: The Wright Brothers and the Process of Invention*. Washington, D.C.: Smithsonian Institution Press.

Jane's Information Service. n.d. *Jane's All the World's Aircraft*. http://jawa.janes.com/docs/jawa/search.jsp. Accessed March 31, 2008.

Jex, H. R. n.d. "Making Pterodactyls Fly (QN Story)." http://www.twitt.org/QNStory.html. Accessed May 12, 2008.

Jones, G., and J. Rydell. 2003. "Hawking and Gleaning." In *Bat Ecology*, edited by T. H. Kunz and M. B. Fenton, pp. 301–45. Chicago: University of Chicago Press.

Jones, R. M. 1999. *Mechanics of Composite Materials*. 2d ed. Philadelphia: Taylor and Francis.

Keennon, M. T., and J. M. Grasmeyer. 2003. "Development of the Black Widow and Microbat MAVs and a Vision of the Future of MAV Design." In *AIAA/ICAS International Air and Space Symposium and Exposition: The Next 100 Years*, pp. 1–11. Dayton, Ohio: American Institute of Aeronautics and Astronautics.

Kershner, W. K. 1979. *The Student Pilot's Flight Manual: From First Flight to the Private Certificate*. 5th ed. Ames: Iowa State University Press.

King, A. S., and J. McLelland. 1984. *Birds: Their Structure and Function*. 2d ed. Philadelphia: Bailliáere Tindall.

Kirkpatrick, S. J. 1994. "Scale Effects on the Stresses and Safety Factors in the Wing Bones of Birds and Bats." *Journal of Experimental Biology* 190, no. 1: 195–215.

Kunz, T. H., and M. B. Fenton. 2003. *Bat Ecology*. Chicago: University of Chicago Press.

Kyle, P. D., and G. Z. Kyle. 2005. *Chimney Swifts: America's Mysterious Birds above the Fireplace*. Louise Lindsey Merrick Natural Environment Series, no. 37. College Station: Texas A&M University Press.

Lack, D. L. 1956. *Swifts in a Tower*. London: Methuen.

Lange, B. 1986. *Typenhandbuch der Deutschen Luftfahrttechnik*. Koblenz, Germany: Bernard and Graefe Verlag.

Langford, J. S. 1988. "Triumph of *Daedalus*." *National Geographic* 174, no. 2: 191–99.

Langston, W., Jr. 1981. "Pterosaurs." *Scientific American* 224: 122–36.

Lehrer, M. 1998. "Looking All Around: Honeybees Use Different Cues in Different Eye Regions." *Journal of Experimental Biology* 201, no. 24: 3275–92.

Leishman, J. G. 2000. *Principles of Helicopter Aerodynamics*. Vol. 12, Cambridge Aerospace Series. New York: Cambridge University Press.

Lilienthal, O. 2001 [1889]. *Birdflight As the Basis of Aviation*, translated by A. W. Isenthal. Hummelstown, Penn.: Markowski International.

Lissaman, P. B. S. 1983. "Low-Reynolds-Number Airfoils." *Annual Review of Fluid Mechanics* 15: 223–39.

Lorell, M. A., and H. P. Levaux. 1998. *The Cutting Edge: A Half-Century of Fighter Aircraft R&D*. Santa Monica, Calif.: RAND.

MacCready, P. B. 1985. "The Great Pterodactyl Project." *Engineering and Science* 49: 18–24.

———. 1986. "Art and Nature in the Technology of Flight." *Leonardo* 19, no. 4: 317–19.

Marullo, C., J. G. Pilon, and M. Mouze. 1987. "Postembryonic Development of the Optic Lobe in *Lestes eurinus* Say and *Aeshna mixta* Latreille: Volumetric Growth (Zygoptera: Lestidae; Anisoptera: Aeshnidae)." *Odonatologica* 16, no. 4: 379–84.

Maynard Smith, J. 1952. "The Importance of the Nervous System in the Evolution of Animal Flight." *Evolution* 6: 127–29.

McMahon, T. A. 1984. *Muscles, Reflexes, and Locomotion*. Princeton, N.J.: Princeton University Press.

Michelson, R. n.d. "Entomopter Project." http://avdil.gtri.gatech.edu/RCM/RCM/Entomopter/EntomopterProject.html. Accessed March 31, 2008.

Miller, G. 2005. "Society for Neuroscience Meeting: Bats Have a Feel for Flight." *Science* 310, no. 5,752: 1260–61.

Miller, L. A., and A. Surlykke. 2001. "How Some Insects Detect and Avoid Being Eaten by Bats: Tactics and Countertactics of Prey and Predator." *Bioscience* 51, no. 7: 570–81.

Milstein, M. 2005. "Superduperjumbo." *Air and Space Smithsonian* 21, no. 2: 22–27.

Mittelstaedt, H. 1950. "Physiologie des Glechgewichtssinnes Bei Fliegenden Libellen." *Zeitschrift für Vergleichende Physiologie* 32: 422–63.

Mohl, B., and L. A. Miller. 1976. "Ultrasonic Clicks Produced by the Peacock Butterfly: A Possible Bat Repellent Mechanism." *Journal of Experimental Biology* 64, no. 3: 639–44.

Moody, P. A. 1970. *Introduction to Evolution.* 3d ed. New York: Harper and Row.

Nachtigall, W. 1974. *Insects in Flight: A Glimpse behind the Scenes in Biophysical Research.* New York: McGraw-Hill.

National Highway Traffic Safety Administration and American Medical Association Commission on Emergency Medical Services. 1986. *Air Ambulance Guidelines.* Revised edition. Washington, D.C.: U.S. Department of Transportation and American Medical Association.

Nickel, K., and M. Wohlfahrt. 1994. *Tailless Aircraft in Theory and Practice.* AIAA Education Series. Washington, D.C.: American Institute of Aeronautics and Astronautics.

Nielsen, J. 2007. "Giant Bats Snatch Birds from Night Sky." *Morning Edition.* Washington, D.C.: National Public Radio, April 18.

Noland, D. 2008. "The Bone Is Back." *Air and Space Smithsonian* 23, no. 1: 60–67.

Norberg, R. Å. 1975. "Hovering Flight of the Dragonfly *Aeschna juncea* L., Kinematics and Aerodynamics." In *Swimming and Flying in Nature,* edited by T. Y. Wu, C. J. Brokaw and C. Brennan, pp. 763–81. New York: Plenum.

———. 1985. "Function of Vane Asymmetry and Shaft Curvature in Bird Flight Feathers; Inferences on Flight Ability of *Archaeopteryx.*" In *The Beginnings of Birds,* edited by M. K. Heckt, J. H. Ostrom, G. Viohl and P. Wellnhofer, pp. 303–18. Eichstätt, Germany: Freunde des Jura-Museums Eichstätt.

Norberg, U. M. 1976. "Aerodynamics of Hovering Flight in the Long-Eared Bat *Plecotus auritus.*" *Journal of Experimental Biology* 65, no. 2: 459–70.

———. 1990. *Vertebrate Flight: Mechanics, Physiology, Morphology, Ecology and Evolution.* Vol. 27 of *Zoophysiology.* New York: Springer-Verlag.

Osorno, J. L., R. Torres, and C. Macias Garcia. 1992. "Kleptoparasitic Behavior of the Magnificent Frigatebird: Sex Bias and Success." *Condor* 94, no. 3: 692–98.

Ostrom, J. H. 1979. "Bird Flight: How Did It Begin?" *American Scientist* 67: 46–56.

Padian, K., and J. M. V. Rayner. 1993. "The Wings of Pterosaurs." *American Journal of Science* 293A: 91–166.

Pennycuick, C. J. 1972. *Animal Flight.* London: Arnold.

———. 1975. "Mechanics of Flight." In *Avian Biology,* edited by D. S. Farner and J. R. King, pp. 1–75. New York: Academic Press.

Pennycuick, C. J., T. Alerstam, and B. Larsson. 1979. "Soaring Migration of the Common Crane *Grus grus* Observed by Radar and from an Aircraft." *Ornis Scandinavica* 10: 241–51.

Popa-Lisseanu, A. G., A. Delgado-Huertas, M. G. Forero, A. Rodríguez, R. Arlettaz, and C. Ibáñez. 2007. "Bats' Conquest of a Formidable Foraging Niche: The Myriads of Nocturnally Migrating Songbirds." *PLoS ONE* 2, no. 2: E205.

Poulin, R. G., S. D. Grindal, and R. M. Brigham. 1996. *Common Nighthawk: "Chordeiles minor."* Birds of North America Series, no. 213. Washington, D.C.: American Ornithologists' Union and Academy of Natural Sciences.

Pringle, J. W. S. 1948. "The Gyroscopic Mechanism of the Halteres of Diptera." *Philosophical Transactions of the Royal Society of London B* 233: 347–84.

Private Pilot Manual. 1981. Denver: Jeppesen-Sanderson.

Prum, R. O. 1998. "Sexual Selection and the Evolution of Mechanical Sound Production in Manakins (Aves: Pipridae)." *Animal Behaviour* 55, no. 4: 977–94.

Raikow, R. J. 1985. "Locomotor Systems." In *Form and Function in Birds*, edited by A. S. King and J. McLelland, pp. 57–148. New York: Academic Press.

Ratcliffe, D. A. 1993. *The Peregrine Falcon*. 2d ed. London: Poyser.

Rayner, J. M. V. 1987. "The Mechanics of Flapping Flight in Bats." In *Recent Advances in the Study of Bats*, edited by M. B. Fenton, P. Racey and J. M. V. Rayner, pp. 23–42. Cambridge, U.K.: Cambridge University Press.

———. 1988. "The Evolution of Vertebrate Flight." *Biological Journal of the Linnean Society* 34: 269–87.

———. 1988. "Form and Function of Avian Flight." *Current Ornithology* 5: 1–66.

———. 1999. "Estimating Power Curves of Flying Vertebrates." *Journal of Experimental Biology* 202, no. 23: 3449–61.

Reay, D. A. 1977. *The History of Man-Powered Flight*. New York: Pergamon.

Rendall, D. 1996. *Jane's Aircraft Recognition Guide*. Glasgow: HarperCollins.

Richarz, K., and A. Limbrunner. 1993. *The World of Bats: The Flying Goblins of the Night*. Neptune City, N.J.: T.F.H. Publications.

Riskin, D. K., S. Parsons, W. A. Schutt, Jr., G. G. Carter, and J. W. Hermanson. 2006. "Terrestrial Locomotion of the New Zealand Short-Tailed Bat *Mystacina tuberculata* and the Common Vampire Bat *Desmodus rotundus*." *Journal of Experimental Biology* 209, no. 9: 1725–36.

Robertson, R. J., B. J. Stutchbury, and R. R. Cohen. 1992. *Tree Swallow: "Tachycineta bicolor."* Birds of North America Series, no. 11. Washington, D.C.: American Ornithologists' Union and Academy of Natural Sciences.

Roper, C. n.d. "Human Powered Flying: Daedalus." http://www.humanpoweredflying.propdesigner.co.uk/html/daedalus.html. Accessed March 31, 2008.

———. n.d. "Human Powered Flying: Jupiter." http://www.humanpoweredflying.propdesigner.co.uk/html/jupiter.html. Accessed March 31, 2008.

———. n.d. "Human Powered Flying: Kremer Speed." http://www.humanpoweredflying.propdesigner.co.uk/html/kremer_speed.html. Accessed March 31, 2008.

———. n.d. "Human Powered Flying: Other '70's Planes." http://www.humanpoweredflying.propdesigner.co.uk/html/other__70_s_planes.html. Accessed March 31, 2008.

———. n.d. "Human Powered Flying: The Gossamers and Other Planes." http://www.humanpoweredflying.propdesigner.co.uk/html/the_gossamers_and_other_planes.html. Accessed March 31, 2008.

———. n.d. "Human Powered Flying: True Flights." http://www.humanpoweredflying.propdesigner.co.uk/html/flights.html. Accessed March 31, 2008.

Roskam, J. 1979. *Airplane Flight Dynamics and Automatic Flight Controls*. Vol. 1. Ottawa, Kans.: Roskam Aviation and Engineering.

Roskam, J., and C. E. Lan. 1997. *Airplane Aerodynamics and Performance*. Lawrence, Kans.: DAR.

Ross, F. K. 1953. *Flying Windmills: The Story of the Helicopter*. New York: Lee and Shephard.

Schmidt-Nielsen, K. 1984. *Scaling: Why Is Animal Size So Important?* New York: Cambridge University Press.

———. 1997. *Animal Physiology: Adaptation and Environment*. 5th ed. Cambridge, U.K.: Cambridge University Press.

Scholey, K. 1986. "The Evolution of Flight in Bats." In *Bat Flight—Fledermausflug*, edited by W. Nachtigall, pp. 1–12. Stuttgart, Germany: Fischer.

Seddon, J. 1990. *Basic Helicopter Aerodynamics*. AIAA Education Series. Washington, D.C.: American Institute of Aeronautics and Astronautics and Blackwell Scientific Publications.

Sherwin, K. 1975. *Man-Powered Flight*. Revised edition. Kings Langley, Hertfordshire, U.K.: Model and Allied Publications.

Shipman, P. 1998. *Taking Wing: Archaeopteryx and the Evolution of Bird Flight.* New York: Simon and Schuster.

Simons, M. 1987. *Model Aircraft Aerodynamics.* Hemel Hempstead, Hertfordshire, U.K.: Argus.

Smithsonian National Air and Space Museum. n.d. "Piper J-3 Cub." http://collections. nasm.si.edu/code/emuseum.asp?profile=objects&newstyle=single&quicksearch= A19771128000. Accessed April 16, 2008.

Spence, C. F. 2002. *Aeronautical Information Manual/Federal Aviation Regulations: AIM/FAR.* New York: McGraw-Hill.

Spenser, J. P. 1998. *Whirlybirds: A History of the U.S. Helicopter Pioneers.* Seattle: University of Washington Press.

Srinivasan, M. V., and S. W. Zhang. 2004. "Visual Motor Computations in Insects." *Annual Review of Neuroscience* 27: 679–96.

Srinivasan, M. V., S. W. Zhang, and N. J. Bidwell. 1997. "Visually Mediated Odometry in Honeybees." *Journal of Experimental Biology* 200, no. 19: 2513–22.

Sternstein, E., and T. Gold. 1991. *From Takeoff to Landing: Everything You Wanted to Know about Airplanes but Had No One to Ask.* New York: Pocket Books.

Swartz, S. M., M. S. Groves, H. D. Kim, and W. R. Walsh. 1996. "Mechanical Properties of Bat Wing Membrane Skin." *Journal of Zoology* 239, no. 2: 357–78.

Tennekes, H. 1996. *The Simple Science of Flight: From Insects to Jumbo Jets.* Cambridge, Mass.: MIT Press.

Thomas, A. L. R. 1993. "On the Aerodynamics of Birds' Tails." *Philosophical Transactions of the Royal Society of London B* 340: 361–80.

———. 1996. "The Flight of Birds That Have Wings and a Tail: Variable Geometry Expands the Envelope of Flight Performance." *Journal of Theoretical Biology* 183: 237–46.

Thomas, A. L. R., and G. K. Taylor. 2001. "Animal Flight Dynamics. I. Stability in Gliding Flight." *Journal of Theoretical Biology* 212, no. 3: 399–424.

Tucker, V. A. 1968. "Respiratory Exchange and Evaporative Water Loss in the Flying Budgerigar." *Journal of Experimental Biology* 48: 67–87.

———. 1993. "Gliding Birds: Reduction of Induced Drag by Wing Tip Slots between the Primary Feathers." *Journal of Experimental Biology* 180: 285–310.

———. 1995. "Drag Reduction by Wing Tip Slots in a Gliding Harris' Hawk, *Parabuteo unicinctus.*" *Journal of Experimental Biology* 198: 775–81.

Tucker, V. A., T. J. Cade, and A. E. Tucker. 1998. "Diving Speeds and Angles of a Gyrfalcon (*Falco rusticolus*)." *Journal of Experimental Biology* 201, no. 13: 2061–70.

Tucker, V. A., and G. C. Parrott. 1970. "Aerodynamics of Gliding Flight in a Falcon and Other Birds." *Journal of Experimental Biology* 52: 345–67.

U.S. Code of Federal Regulations. n.d. Title 14, parts 23 and 25. http://ecfr.gpoaccess.gov. Accessed August 30, 2007.

U.S. Congress, Senate Committee on Armed Services. 2002. *Report of the Panel to Review the V-22 Program: Hearing before the Committee on Armed Services, United States Senate, One Hundred Seventh Congress, First Session, May 1, 2001.* Washington, D.C.: U.S. Government Printing Office.

Van Essen, D. C., C. H. Anderson, and D. J. Felleman. 1992. "Information Processing in the Primate Visual System: An Integrated Systems Perspective." *Science* 255, no. 5,043: 419–23.

Vogel, S. 1967. "Flight in *Drosophila.* III. Aerodynamic Characteristics of Fly Wings and Wing Models." *Journal of Experimental Biology* 46: 431–43.

———. 1988. *Life's Devices: The Physical World of Animals and Plants.* Princeton, N.J.: Princeton University Press.

———. 1994. *Life in Moving Fluids: The Physical Biology of Flow.* 2d ed. Princeton, N.J.: Princeton University Press.

———. 2001. *Prime Mover: A Natural History of Muscle*. New York: Norton.

Voigt, C. C., and Y. Winter. 1999. "Energetic Cost of Hovering Flight in Nectar-Feeding Bats (Phyllostomidae: Glossophaginae) and Its Scaling in Moths, Birds and Bats." *Journal of Comparative Physiology B: Biochemical Systemic and Environmental Physiology* 169, no. 1: 38–48.

von Kármán, T. 1963 [1954]. *Aerodynamics*. Reprint. New York: McGraw-Hill.

Wang, J. n.d. "Gyros for R/C Helicopters." http://www.heliproz.com/jwgyros.html. Accessed March 31, 2008.

Wang, Z. J. 2004. "The Role of Drag in Insect Hovering." *Journal of Experimental Biology* 207, no. 23: 4147–55.

Ward, S., U. Moeller, J. M. V. Rayner, D. M. Jackson, D. Bilo, W. Nachtigall, and J. R. Speakman. 2001. "Metabolic Power, Mechanical Power and Efficiency during Wind Tunnel Flight by the European Starling *Sturnus vulgaris*." *Journal of Experimental Biology* 204, no. 19: 3311–22.

Ward-Smith, A. J. 1984. *Biophysical Aerodynamics and the Natural Environment*. New York: Wiley.

Weis-Fogh, T. 1956. "Biology and Physics of Locust Flight. IV. Notes on Sensory Mechanisms in Locust Flight." *Philosophical Transactions of the Royal Society of London B* 239, no. 667: 553–84.

Whitman, D. W., L. Orsak, and E. Greene. 1988. "Spider Mimicry in Fruit Flies (Diptera: Tephritidae): Further Experiments on the Deterrence of Jumping Spiders (Araneae: Salticidae) by *Zonosema vittigera* (Coquillett)." *Annals of the Entomological Society of America* 81, no. 3: 532–36.

Wilcox, R. K. 1990. *Scream of Eagles: The Creation of Top Gun and the U.S. Air Victory in Vietnam*. New York: Wiley.

Wilkie, D. R. 1960. "Man As an Aero Engine." *Journal of the Royal Aeronautical Society* 64, no. 596: 477–81.

———. 1960. "Man As a Source of Mechanical Power." *Ergonomics* 3, no. 1: 1–8.

———. 1985. "Muscle Function: A Personal View." *Journal of Experimental Biology* 115: 1–13.

Williams, M. 2003. "The 156-Tonne Gimli Glider." *Flight Safety Australia* 7, no. 3: 22–27.

Wills, P. A. 1960. *The Beauty of Gliding*. New York: Pitman.

Wiltschko, W., and R. Wiltschko. 1972. "Magnetic Compass of European Robins." *Science* 176: 62–64.

———. 1988. "Magnetic Orientation in Birds." In *Current Ornithology*, edited by R. F. Johnston, pp. 67–121. New York: Plenum.

Winston, M. L. 1987. *The Biology of the Honey Bee*. Cambridge, Mass.: Harvard University Press.

Withers, P. C. 1981. "An Aerodynamic Analysis of Bird Wings As Fixed Aerofoils." *Journal of Experimental Biology* 90, no. 1: 143–62.

Wood, R. J., S. Avadhanula, M. Menon, and R. S. Fearing. 2003. "Microrobotics Using Composite Materials: The Micromechanical Flying Insect Thorax." In *Proceedings of the IEEE International Conference on Robotics and Automation*, pp. 1842–49. Piscataway, NJ: IEEE Press.

Wooldridge, E. T. 1983. *Winged Wonders: The Story of the Flying Wings*. Washington, D.C.: Smithsonian Institution Press.

———. 1995. *Into the Jet Age : Conflict and Change in Naval Aviation, 1945–1975: An Oral History*. Annapolis, Md.: Naval Institute Press.

Wortman, M. 2006. *The Millionaires' Unit: The Aristocratic Flyboys Who Fought the Great War and Invented American Airpower*. 1st ed. New York: PublicAffairs.

INDEX

adverse yaw, 70, 83; birds and bats, 91; correcting with rudder, 71; Gossamer Condor turns using, 214

aerial mating: dragonflies, 179; forced, 179; hover flies, 180; swifts, 177

aerial predation: bats hawking insects, 175–176; bats, on migrating birds, 174–175; birds, upon insects, 176; coevolution of predator and prey, 173, 184; comparison with air combat, 182; falcons and other birds, 173; frigatebird, 173; maneuverability as defense, 184; rarity among species, 171; required skills, 170; robber flies (asilids), 178; size differences, 175; vision, 185

aerial predators: evasion of, 172; falcon-style stoop, 171; size and speed, 171; speed and maneuverability, 191

aeronautical charts, 107, 119

AeroVironment, 222, 225, 227

agility. See maneuverability

ailerons, 58, 72, 81; bird tails, 93; history, 72; spoiler type, 73

Airbus A320, 1

Airbus A380, 19, 61

airfoil, 58; camber, 27; defined, 25; helicopter rotor, 147; metal ribs and, 31; movement of center of lift, 95; reflexed, 95; sailplanes, 140; symmetrical (no camber), 28

Air Force Academy, 124

airliners: Airbus A320, 1; Airbus A340, 60; Airbus A380, 19; aspect ratio, 19; autopilot, autoland function, 111; Boeing 707, 60; Boeing 727, 60; Boeing 737, 1, 60; Boeing 747, 60, 228, 232; Boeing 767 gliding, 42;

ceilings, 138; Concorde, 98; DC-9/MD-80, 59; fly-by-wire controls, 80; Fokker F-28, 28; gliding without power, 42; ground proximity alarm, 121; terminal space limits wingspan, 19; turboprop-powered, 57; typical flight, 1–2; upper size limits, 61; wing loading, 30

airplanes: aerobatic (Pitts Special), 28; aerobatic, hovering, 164; camber of wing, 27; ceilings, 138; crosswind landings, 89; efficiency compared with flying animals, 233; materials used to build, 24–25; military, in World War I, 181; motorglider, 126; Piper Cub, 126; R/C model wing loading, 30; radar, 105; radio control, 61; tail, 83; tailless, 94–96, 98–99; upper size limits, 61

airplane structural materials, 24–25

airplane tails: as stabilizers, 84; drag on, 95; preventing angle of attack changes, 95

airplane wings: fiberglass, 32, 141; internal structures, 31; metal, 31; structural materials, 31–33; uses other than flight, 37; wood and fabric, 31

airports: altimeters and elevation, 121; Category IIIc (no visibility) landings at, 111; gates limit wingspans, 19; helicopter airlines to, 154; runways and, 142

airspeed: compressibility, 190; dragonflies, 178; falcon, 172; falcon, diving, 192; Harrier jump-jet, 165; helicopters, maximum, 162; inaccurate bot fly estimate, 192; insects, 192; predator versus prey, 172; tiltrotor maximum, 163; versus ground speed, 117

air-to-air combat, 183, 185; analogy to predation, 170; comparison with aerial predation, 182;

ABOUT THE AUTHOR

David E. Alexander grew up outside Dayton, Ohio, about twenty miles from the boyhood home of Orville and Wilbur Wright and a similar distance from the U.S. Air Force Museum. After earning a B.S. from the University of Michigan and a Ph.D. from Duke University, he taught biology for several years at Bellarmine College in Louisville, Kentucky. He currently teaches at the University of Kansas, where he has studied the biomechanics of animal swimming and animal flight. He has had a lifelong passion for anything that flies; and when he is not studying animal flight, he is reading about aviation history and technology or building airplane models. He enjoys traveling with his family, flying radio-control airplanes, snorkeling and scuba diving, and boating.